MARY AND BIOETHICS:
AN EXPLORATION

FRANCIS ETHEREDGE

En Route Books and Media, LLC
St. Louis, MO

Make the time

En Route Books and Media, LLC
5705 Rhodes Avenue
St. Louis, MO 63109

Cover credit: TJ Burdick

Library of Congress Control Number: 2020944499

Copyright © 2020 Francis Etheridge
All rights reserved.

ISBN-13: 978-1-952464-22-5

No part of this booklet may be reproduced, stored in a retrieval system, or transmitted in any form, or by any means, electronic, mechanical, photocopying, or otherwise, without the prior written permission of the author.

Contents

Acknowledgements v

General Foreword by Dr. Anthony Williams 1

Prologue: Pain in the "Biological Family"; "Soma: a Holiday from Reality"; and "Mary and Bioethics" 5

General Introductions to Each Chapter of the Book: Chapters One to Seven... 19

Foreword to Chapter 1: Dr. Mary Anne Urlakis 33

Chapter One: The Holy Family: Celibacy and Marriage: A Reflection on the "Passage" from the Jewish Rite of Marriage to the Christian Sacrament of Marriage: 49

General Introduction: Background: Continuity between the Old and the New Covenant; A new Appreciation of the Transformation of Marriage; Background: Scripture and Marriage

The Book of Tobit (I); Christ and the Covenant (II); The Marriage of Mary and Joseph (III); Christ and His vocation to Celibacy (IV): Cornelius and his family (V); Conclusion.

Foreword to Chapter 2: Maria McFadden Maffucci 85

Chapter Two: Part I of a Marian Triptych: Mary is the Choice of God: 89

My interest in whether or not there needs to be a new Dogmatic Statement on the Mystery of Mary (I); Mary 'Gate of Heaven' (II); Sharers in the Suffering of Christ (III); Some Objections to this Point of View (IV); and Conclusion (V).

Foreword to Chapter 3: Laura Elm 105

Chapter Three: Part II of a Marian Triptych: Our Hope in Mary: 111

Hope in the Reality of the Holy Family (I); The Definition of Hope and the Coming of Christ (II); How many Hopes Appear to us as Impossible?! (III); How Does Mary Help us to Hope? (IV); Two Objections to our Hope in Mary (V); Conclusion.

Foreword to Chapter 4: Edmund Adamus 127

Chapter Four: Part III of a Marian Triptych: Mary and Prayer: 131

The Emergence of Mary (I); Mary and the Prayer of the Church (II); Marian Prayer and the Covenant: The Conversation of Conversion (III)

Foreword to Chapter 5: Dr. Michal Pruski 145

Chapter Five: The First Instant of Mary's Ensoulment: 151

Introduction: Two Possible Meanings of the One Reality of Human Conception; The Dogma of the *Immaculate Conception*: Two Instants or One Moment? (I); The Help of the Doctrine of Original Sin (II); A Discrepancy Between the Implied First Instant of Fertilization in the Dogma of the *Immaculate Conception* and the Definition of Conception in the English Translation of *Donum Vitae* (III); Conclusion: The Need for a More Precise Definition of Human Conception in the Documents of the Church

Foreword to Chapter 6: Dr. Moira McQueen 175

Chapter Six: Mary and Bioethics: 193

Introduction: Being-in-Relationship; Man, male and female, Christ and Mary (I); Mary and Bioethics (II); Mary and Gender Ideology (III); Marriage and Parenting: A Dynamic Conclusion

Contents

Foreword to Chapter 7: Leah Palmer 223

Chapter Seven: Love, Scripture, Suffering and Bioethical Questions: 229

The command to "Be fruitful and multiply" and the Cross of Infertility (I); The Biblical Account of the Suffering of Infertility (II); and, finally, Love and Bioethics (III)

Epilogue: 255

The manufacture of children and the reality of relationship; Biblical prophecy and the Church's criticism of fathers; Is there a way back to the future of marriage and family life?; Reasonable and graced humility; The radical originality of God

An End Word: A New Beginning – by Bishop John Keenan of Paisley, Scotland 267

Acknowledgments

The more I ponder the controversies and developments that characterize bioethics the more the theme of suffering arises: not only in the circumstances of many lives that have driven people to seek remedies but often in the remedies themselves there is a further outbreak of suffering. Therefore, rather than place this quotation at the end or at any other place in this book it is fitting to put it at the beginning.

Fr. Joseph Tham says, in his reflection on Job: 'Even though suffering universally touches our profoundest sensibilities and provokes in us a yearning for answers about our origin, purpose, and end, it is rarely a subject of interest in medical or bioethics journals. It is unfortunate that contemporary bioethics, because of historical reasons, has excluded a deeper discussion on suffering, possibly because of its secular bias (Evans, 2002; Tham, 2008)'[1].

At the same time, however, we live in the time between the coming of Christ and the Second Coming; and, therefore, we live in the favourable moment of the redemption as Pope Francis says: 'Every detail of the life of the body and of the soul, in which the love and redemptive power of the new creation shine forth within us, leads to amazement before the miracle of a resurrection in the very process of occurring (cf. *Col* 3:1-2)' (*Humana Communitas*, 13); and, as such, we all need the help of a mother: 'Today we also need a mother. So we entrust to the maternal love of Mary, Our Lady of the Way, of so many painful journeys, all migrants and refugees, together

---

[1] Fr. Joseph Tham, p. 2 of 18, "Communicating with Sufferers: Lessons from the Book of Job", Christian Bioethics Advance Access published March 22, 2013: *Christian Bioethics:* doi:10.1093/cb/cbt003.

with those who live on the peripheries of our world and those who have chosen to share their journey'[2].

In view of the many difficulties of life that we all face let us invoke the help of the Holy Family and that great crowd of witnesses (Heb. 12: 1) which have gone before us beginning with 'St. Matilda: A Marian saint and patroness of large families'[3]. At the same time we need a spirituality which calls on the healing of the whole heart; and, as this book has developed, so I hope that 'heart speaks to heart'[4] of the Lord's reconciling, redemptive love of us all.

Even if I am impatient of the completion of a book, I believe that the benefit which arise out of these delays outweighs the drawbacks and this work is no exception as there has arisen further considerations. On the one hand, Pope Francis speaks of "being 'unsettled by the living and effective word of the risen Lord[5]'"; and, therefore, he is speaking of the "word of the risen Lord" as a unique word which can pass us to where we cannot take

---

[2] 'Today We Also Need a Mother,' Pope Appeals to Entrust Painful Journeys of All on the Move to Mary (Full Homily for World Day of Migrants), 29th September, 2019, Vatican translations: https://zenit.org/articles/today-we-also-need-a-mother-pope-appeals-to-entrust-painful-journeys-of-all-on-the-move-to-mary-full-homily/?

[3] The name and title of this saint was sent to me by "A Moment with Mary": https://us3.campaign-archive.com/?e=83d33a4ae4&u=bbaf519c73482457368060b5b&id=169401e45b, who drew on the following website for its information: https://www.catholicnewsagency.com/saint/st-matilda-177.

[4] The motto St. John Henry Newman took on becoming a cardinal; see p. 111 of Fr. Ian Ker's, *Healing the Wound of Humanity: The Spirituality of John Henry Newman*, London: Darton, Longman and Todd, 1993.

[5] https://www.hprweb.com/2019/09/letter-of-his-holiness-pope-francis-to-priests/#fnref-24676-26.

ourselves without His help[6]. But, on the other hand, as Fr. John O'Brien says in *A Love Supreme*[7]: 'The trials and pains of the world can silence this voice for a while. We are never far from the grace of God. He continues to reach out to touch us. We have to become sensitive again to his voice'[8]. It is true, too, that there are a variety of experiences in which a person encounters almost a bleakness in front of the difficulties of life; and, therefore, in that desert there are many possibilities – not least is the search for the meaning of the event: a search which may change but not cease[9]: a search in which we may need to be sustained and assisted if we are not to abandon it to the detriment of imperilling the life lived without adequate answers.

You will discover, as I have when reading the various contributions, a number of beautiful, even tearfully beautiful, faith-enriched experiences amidst all that these contributing authors bring to this book; and, in general, it is true that the experience of suffering touches our heart deeply and, possibly more than anything else, raises the questions that need answers.

As a summary, however, of the fruit of faith, let us say:

God comes to meet us where we are, loves us as we are and takes us where we cannot go without Him.

So, as we read and pray through this book, let us hope in a renewed sensitivity to the help of the word of God in that search which draws on all that is human and needs the help of all that is of God.

---

[6] Etheredge, Francis: "Conscience as Relationship: Part II": https://www.hprweb.com/2020/02/conscience-as-relationship-part-ii/.

[7] From an original text supplied by the author, John O'Brien, OFM (Franciscan), p. 53, otherwise published under the same title on Amazon: https://www.amazon.co.uk/Love-Supreme-John-Obrien OFM/dp/1687860106.

[8] From an original text supplied by the author, John O'Brien, OFM (Franciscan), p. 53, otherwise published under the same title on Amazon: https://www.amazon.co.uk/Love-Supreme-John-Obrien OFM/dp/1687860106.

[9] Cf. Viktor E. Frankl, *Man's Search for Meaning: The Classic Tribute to Hope from the Holocaust*, published in 2004 by Rider, an imprint of Ebury Publishing.

Finally, having drafted the book, I would like to thank all those who have contributed so thoughtfully to the completed work – enriching it so well with a mixture of learning, discernment and invaluably encouraging experience: Dr. Anthony Williams for his General Foreword; the following who have each contributed a Foreword to one of the seven Chapters, so well and so varied: Dr. Mary Anne Urlakis; Maria McFadden Maffucci; Laura Elm; Edmund Adamus; Dr. Michal Pruski; Dr. Moira McQueen; Leah Palmer; and, finally, an End Word: A New Beginning, by Bishop John Keenan of Paisley, Scotland.

GENERAL FOREWORD

DR. ANTHONY WILLIAMS

Mary and Bioethics: An Exploration will be found to contain very valuable information for student, researcher and interested reader. This is an important book on a subject now calling for profound reflection by everyone for whom the Motherhood of Mary is a central reality in life. The themes it pursues could not be of greater immediacy for the life of the human race in the 21st century.

A great African bishop, Cardinal Robert Sarah, recently made the following prophetic observation about the state of the Western world today:

> "The West refuses to *receive*, and will accept only what it constructs for itself … Because it is a gift from God, human nature itself becomes unbearable for Western man. This revolt is spiritual at root. It is the revolt of Satan against the gift of grace. Fundamentally, I believe that Western man refuses to be saved by God's mercy. He refuses to receive salvation, wanting to build it for himself" (From his book, *The Day is Far Spent*, Ignatius Press).

Francis Etheredge invites us "to consider anew the very first thing, which is not what do we think, but what has God done?"[1]. The very first things are Creation, which is the work of the Divine Word, and Incarnation, which

---

[1] In Chapter Two: Part IV: Mary Gate of Blessing.

could not have happened without the co-operation of Mary, the Mother of the Word. A world which refuses to receive these realities as a gift from God begins to find human nature unbearable, and something to be re-created in the image of whoever has the power to redefine it and impose the redefinition.

The author takes us back through the history of Israel and the instances of sterility, in the lives of Sarah, Rachel and Hannah, which opened the way for what "God has done" in the history of salvation. Finally, through the apparent "sterility" of the virginity of Mary, came the Saviour, in whom the whole meaning of life is made manifest. Life is in the first place a gift, and only acceptance of the gift leads to true understanding of human nature, and to eternal life. In this respect, "the identity of Mary is like a glorious secret within the life of the Church"[2].

Chapter five takes us to heart of the author's concern: the significance of the doctrine of the Immaculate Conception of Mary, our mother in the Lord, for a true understanding of the relationship between conception and ensoulment in every unborn human child. Does the Immaculate Conception carry the final confirmation that the two moments, of conception and ensoulment, are one and the same? The author's answer is a clear 'Yes', an answer which has resounding implications not only for the dignity of the unborn child, but for acceptance of life as a gift. In this way, Mary can be seen, in the author's words, as "the perfect exponent of our own beginning". This is linked to the Church's documents on both these issues, together with an appeal for further consideration and clarification.

Francis shows how ingratitude, rejection of life as a gift, leads to rejection of suffering as a redemptive reality, and to an angry determination to use every means to avoid it, by the use and misuse of science. In this way we never come to know either who we are, or the glory for which were created and to which we are called. The author, in his own inimitable style, calls on

---

[2] In Chapter Three: Part IV: How Does Mary Help us to Hope?

General Foreword

his own life experience to centre the message firmly in the daily reality of our calling. The result might take the reader by surprise[3], but can hardly diminish the importance of the overall message.

The author makes use of the work of a militant feminist, "Firestone", to show the deeply destructive implications of a false account of love, and of an ethic which offers "pansexuality" as the meaning and goal of human life and the way to a better world. The outcome can only be hell-on-earth, for children first, and then for adults as well.

Bioethics is able to serve human life fruitfully just so long as human life is accepted as a Divine gift and Divine calling. Once "liberated" from this context, and from the wisdom that goes with it, bioethics can only become the servant of death. In a world in which marriage and the family are being "deconstructed", and humanity reduced to a collection of lonely "singulars", what Francis has to say is both a matter for profound reflection and a call to prayer, conversion and action.

Anthony Williams, M.A. (Theol.), Ph.D.

---

[3] In Chapter Four: Part IIIiii: The Experience of Prayer "Beads".

PROLOGUE

What possible point can there be to a book on *Mary and Bioethics?*[1] Is it not bringing together two different and unconnected subjects and, as it were, scarcely making a connection by their juxtaposition? The book as a whole, it could be argued, is an exercise in apologetics, a defence of Christian and indeed of Marian doctrine; however, this view suggests that responding to what we suffer is almost marginalized if not actually "boxed" and shifted to an obscure place called "religion", "Catholic doctrine" or even "Marian doctrine". The reality, however, is completely different: God's love for sinners comes to us in our concrete situation; and, indeed, *Love* comes to be with us and, in being with us, to take us where Love is. Beginning with the reality of human experience roots the reality of salvation in the history of the human race and in our own history.

Are there not questions that need answers instead of the polemical accusations that the world needs an end to children, or men and women or indeed that the planet needs to exist without us to recover its health? Is it not a controversial title, *Mary and Bioethics*, to attract attention and, therefore, sales? Is it wrong to attract attention to questions which, in one way or another, are all around us; and, therefore, is it wrong to enter the debate while at the same time hoping to earn a living? Why start with *Mary and Bioethics* when there are so many more specific and even specialized bioethical

---

[1] I acknowledge a debt to a brief but helpful correspondence with Prof. Brian K. Reynolds, who helped me to think more widely with respect to a work on Mary (email, 16/1/2019). I am also very grateful to Dr. Anthony Williams for his proof-reading, numerous suggestions and General Foreword.

questions "out there"? But then why not start with basic questions about human identity? Is it not a part of the problem that too many bioethical questions are discussed without foundations as if, as it were, everything is new and there is nothing that existed before the latest "procedure", "application" or "experiment"?

Beginning with Mary is, in a sense, about going back to the beginning of a lot of modern questioning and asking the reason for questioning whether or not anything exists as it does for a good purpose? Certainly it is unjust to women to restrict their employment opportunities to a range of tasks that do not reflect their actual human talents, training and capabilities; but, in reality, is this a reason to ask: Is a man a man, a woman a woman, or a marriage for a man and a woman a good outcome for them or is a family a good outcome of marriage? Thus, one wonders, what has started this questioning of what exists, even if what exists is imperfect; indeed, is it the very imperfection of what exists that troubles the human heart?

Pain in the "biological family"

Into this medley of questions, then, comes a pointed comment by a Marxist-feminist, Shulamith Firestone, who is quoted as writing the following in 1970: the "'end goal of feminist revolution must be … not just the elimination of male *privilege* but of the sex distinction itself." Then "the tyranny of the biological family would be broken," "unobstructed pansexuality" would replace heterosexuality, and "all forms of sexuality would be allowed and indulged." Firestone argued that "[u]nless revolution uproots the basic social organization, the biological family … the tapeworm of exploitation will never be annihilated".[2] Indeed, it is even possible that

---

[2] Excerpt from an article entitled, "The Trans-Industrial Complex" by Mary Hasson, *Humanum: Issues in Family, Culture and Science*, Issue 2: http://humanumreview.com/articles/the-trans-industrial-complex.The article has a link Firestone's book *The Dialectic of Sex: The Case for Feminist*

she saw the problem epitomized, as it were, in a particular kind of small family: 'the nuclear family of a patriarchal society, a form of social organization that intensifies the worst effects of the inequalities in the biological family itself'[3]. In the final expression of this excerpt it is the "biological family" that is "the tapeworm of exploitation" which needs to be "annihilated"; and, in the lead up to this annihilation, the method of which is the destruction of being male or female and of any sexual order whatsoever. It is implied, therefore, that Firestone experienced or witnessed that the "biological family" is "the tapeworm of exploitation"; and, therefore, as a tapeworm is what generally lives in a human intestine, feeding on the food that is intended for the host human being, there is evoked an experience so painful that she wants to eradicate the "biological family" altogether.

Whether it is fair or not to draw this conclusion from limited evidence there is, nevertheless, a passing reference to the 'most rigid' system in existence which, therefore, has helped to determine the 'radical feminist' aim to 'overthrow ... the oldest, most rigid class/caste system in existence, the class system based on sex ... lending the archetypal male and female roles an underserved legitimacy and seeming permanence'[4]. Bearing in mind, then, Firestone's Jewish background[5], is it possible that she experienced or saw a

---

Revolution:https://www.amazon.com/Dialectic-Sex-Case-Feminist-Revolution/dp/0374527873/ref=sr_1_1?ie=UTF8&qid=1536436757&sr=8-1&keywords=shulamith+firestone+the+dialectic+of+sex&dpID=514jT9a7qWL&preST=_SY291_BO1,204,203,200_QL40_&dpSrc=srch

[3] Firestone, *The Dialectic of Sex: The Case for Feminist Revolution*, pp. 43-44 and other references to it elsewhere.

[4] Firestone, *The Dialectic of Sex: The Case for Feminist Revolution*, p. 15.

[5] Brief biography at the beginning of *The Dialectic of Sex: The Case for Feminist Revolution* (no page number); but also note the uncritical reference to creation which argues for a residual willingness to consider that there are deeper causes of human nature than that of culture: 'Unlike economic class, sex class sprang directly from a biological reality: men and women were created different, and not equal' (p. 8).

particularly rigid expression of role divisions based on sex? In her review of the roles and attitudes that characterize fathers and mothers there is an almost stereotypical negativity about the father and the mother. The father 'makes ... [the] mother unhappy, makes her cry, doesn't talk to her very much, argues with her a lot, bullies (this is why, if ... [the son] has seen intercourse, he is likely to interpret it on the basis of what he has already gathered of the relationship: that is, that his father is attacking his mother)'[6]. Children, generally, 'don't plan to be stuck with the lousy limited lives of women'. The mother 'cares for ... [her children] more closely than the father, and shares her oppression with her'[7] [daughter, if not also with her son – but what convinces the son to identify with the father 'is the offer of *the world*

However, it is not clear why being 'created different' is equivalent to not being 'equal'; for, in general, we are all a "gift" and not self-created. Therefore equality is not equivalent to being identical; indeed, if being identical was a precondition for equality – equality would be impossible as real individuality entails a multitude of differences within the reality of being male and female. On p. 67, referring to the Old Testament account of 'Jacob's family train as ... he travels to meet his twin brother Esau'., Firestone says: 'This early patriarchal household was only one of many variations on the patriarchal family taking place in many different cultures up to the present time.' As I shall argue later, in Chapter 7, the problems in the history of this family are echoed in today's exploitation of the suffering of infertility. Cf. also p. 155 and then on p. 197 where Firestone uses a Hebrew expression for a kibbutz, '*Beit Yeladim*' (House of ...).

[6] Firestone, *The Dialectic of Sex: The Case for Feminist Revolution*, p. 47; cf. also p. 63 where she speaks of 'marital hell' and 'the rigid patriarchal family' and p. 65 where she says of women and children that their bond 'is no more than a shared oppression' and the blight of the 'patriarchal nuclear family' etc. At the same time, however, there are softer expressions where she both refers to her father's induction into adulthood via changes of clothing (p. 73) and to her Jewish background as illustrating the advance of children's abilities when they are not trimmed to fit the modern developmental category of children's activities (cf. p. 75).

[7] Firestone, *The Dialectic of Sex: The Case for Feminist Revolution*, p. 48.

when he grows up': 'of travel and adventure'[8]]. In other words there is no indication of the sensitivity of the man to the woman or of the woman's capacities being equivalently developed and appreciated; and, therefore, there is no sense of the remarkable opportunities and accomplishments which have been exemplified precisely because of the religious recognition of the vocation of women; indeed, it could be argued, there is a work to be done in terms of recognizing the graced development of innumerable women in the history of salvation – beginning in the Old Testament and going on through the coming of Christ and the mission of the Church[9]. In a word, conversion, whether of the man or the woman to the help of God is an as yet frequently untold account of the positive contribution of grace to the healing of the dysfunctional effect of original and personal sin on the relationship of men and women.

At the same time, however, the Jewish reality is richer than specific instances or experiences reveal and already, much earlier, in early 19th Century Europe, Edith Stein's mother was taking on and successfully developing a timber business on the death of her husband[10]; and, indeed, Edith Stein was herself both a pioneering student of philosophy and a modern writer on the legitimate development of modern feminism[11].

---

[8] Firestone, *The Dialectic of Sex: The Case for Feminist Revolution*, p. 47.

[9] Cf. Pope St. John Paul II, *Letter to Women*; and, indeed, he acknowledges the downside of this history too, when he says: 'And if objective blame, especially in particular historical contexts, has belonged to not just a few members of the Church, for this I am truly sorry. May this regret be transformed, on the part of the whole Church, into a renewed commitment of fidelity to the Gospel vision' (3).

[10] Cf. Susanne Batzdorff's, *Aunt Edith: The Jewish Heritage of a Catholic Saint*, (Springfield, Illinois: Templegate Publishers, second edition, 2003) p. 75; and cf. also, the essay on Edith Stein in Francis Etheredge, *The Family on Pilgrimage: God Leads Through Dead Ends*, St. Louis: En Route Books and Media, 2018.

[11] Edith Stein's mother encouraged the education of both her very able daughters (cf. Batzdorff's, *Aunt Edith: The Jewish Heritage of a Catholic Saint*, pp. 102-110).

Nevertheless, it is possible that there is a kind of cross-cultural rigidity in role differences between the sexes which is indicative of a psychology of power and not of service; and, therefore, there is truth to the claim that there is a cultural contribution to domination as an exploitation of gender differences[12]. In other words, the problem of "role rigidity" is cross-cultural and requires a deeper remedy than that of "exterior" social practices, however helpful those changes may be in terms of opening up all kinds of legitimate opportunities for both sexes to be more humane in their treatment of each other.

What arises out of all this is the problem of psychological pain: the pain that arises from the experience of being an oppressed woman: 'This is painful: no matter how many levels of consciousness one reaches, the problem goes deeper'[13] until she arrives at the ultimate cause of it all being the sex difference: 'men and women were created different, and not equal'[14]. Indeed, Firestone specifically says: 'I had to train myself out of that phony smile [that 'indicates acquiescence of the victim to her own oppression'] … And this meant that I smiled rarely, for in truth, when it came down to real smiling, I had less to smile about'[15]; and, having already referred multiple times to the rigidity of patriarchy, she says of a type of childhood which falsifies growing up because it is disengaged from adulthood, that a boy will

---

[12] Cf. Excerpt from the "Letter to the Bishops of the Catholic Church on the Collaboration of Men and Women in the Church and in the World", 2: http://www.vatican.va/roman_curia/congregations/cfaith/documents/rc_con_cfaith_doc_20040731_collaboration_en.html: 'In order to avoid the domination of one sex or the other, their differences tend to be denied, viewed as mere effects of historical and cultural conditioning. In this perspective, physical difference, termed *sex*, is minimized, while the purely cultural element, termed *gender*, is emphasized to the maximum and held to be primary.'

[13] Firestone, *The Dialectic of Sex: The Case for Feminist Revolution*, p. 3.

[14] Firestone, *The Dialectic of Sex: The Case for Feminist Revolution*, p. 8.

[15] Firestone, *The Dialectic of Sex: The Case for Feminist Revolution*, p. 81.

grow up 'to become a robot like his father'[16]. Psychological pain can be so profound, then, that if the person cannot respond to it creatively all that arises is the desire to eradicate the "cause" of it. In this case, if the "cause" is the "biological family", then in the very terms of this expression there is already an expression of psychological distance from what has brought about this suffering; and, therefore, a kind objectification of what needs to be got rid of. If a child was not destroyed by the experience of the "biological family" then there would be many references to the parents, grandparents and siblings as brothers and sisters, mom and dad, granny and grandpa (or even more endearing expressions). Thus in seeking an explanation for the desire to destroy what exists there emerges the possibility of the problem of pain: a pain that seeks to destroy what has caused it. Indeed, according to Firestone, it is because the 'biological family' is an inevitable vehicle 'through which the psychology of power can always be smuggled' that it must be destroyed; and, as such, she thinks that she has taken 'the class analysis one step further to its roots in the biological division of the sexes'[17]. In other words, the point arises that if original sin is rejected, a sin disrupting the harmony between man and woman and creation, then it follows that people search for an alternative "origin" or "cause" to the disorders that exist; and, whatever the real merit of the explanation, the danger is that there is a disproportionate emphasis on the symptoms of social disorders because of the want of a primary explanation which bears on the whole human race[18].

Whether it arises out of personal experience or not the goal of the 'annihilation' of the 'biological family' is an almost completely negative ambition to destroy what exists as if, by doing so, it is possible to eradicate the desire for good relationships which is so fundamental to human

---

[16] Firestone, *The Dialectic of Sex: The Case for Feminist Revolution*, p. 83.

[17] Firestone, *The Dialectic of Sex: The Case for Feminist Revolution*, p. 12.

[18] Cf. Chapter 5 of this book on the doctrine of original sin and its usefulness in terms of arriving at an understanding of human conception derived from the Dogma of the Immaculate Conception of Mary, the Mother of the Lord.

happiness. In other words, instead of seeking to heal, remedy or rectify the problems of family life, there is a kind of intensification of the suffering experienced as there is a continual rejection of the good of human relationships; indeed, just as a person frightened by a shadow might want to eradicate all shadows, the reality of seeking to destroy all shadows entails the destruction of all that exists. Recognizing the disfigured reality of family life entails recognizing even more fully the redemptive effect of Christ's redeeming love.

"Soma: a holiday from reality"

In Aldous Huxley's *Brave New World* there is more than an echo of Firestone's rejection of the "biological family". The question is: What was already happening in 1932 that inspired Huxley's account of a possible "farming of human beings in the future"? What was already happening in 1818 that inspired Mary Shelley's account of "constructing" a human being from parts of dead bodies in *Frankenstein*? Indeed, the whole book is almost a warning to would-be-scientists to be wary of "discovery blindness" unleashing harm on the human race[19]. What was already happening in between when H. G. Wells wrote *The Island of Dr. Moreau* about animal-human hybrids in 1896? Again there is a warning that harmful experiments lead to harm; moreover, the experimentalist hides what he is doing or kills what escapes his control[20]. What if those who experiment on human embryos saw the outcome of their experiments in terms of what it did to a person's life? Were all of these pure "leaps of the imagination" or was there, as it were,

---

[19] Mary Shelley, *Frankenstein or The Modern Prometheus*, edited with an Introduction and Notes by Maurice Hindle, revised edition, London: Penguin Books, 1992.

[20] H. G. Wells, *The Island of Doctor Moreau: A Possibility and Other Stories*, with an Introduction and Notes by Emily Alder, Hertfordshire: Wordsworth Classics, 2017.

what prepared the ground for each writer's envisioning of the "future"?[21] What is already happening in today's fiction?

In *Brave New World* there is an early account of the factory "processing" of human beings, from conception until when they were deemed old enough to gravitate from this process and be subject to the psychological as well as the physical procedures, whereby it was hoped to determine what part a person would play in society. There was, then, a robust rejection of the psychological pain, suffering and messiness of family life: 'And home was as squalid psychically as physically': 'What suffocating intimacies, what dangerous, insane, obscene relationships between the members of the family group': 'Maniacally, the mother brooded over her children (*her* children)'[22]. In the way that this family life is described it is possible that there is an element of the dysfunctional in the family life which is, otherwise, structured around husband and wife, parents and children. To emphasize the problematic nature of marriage and family life there is a reservation where the original family order is allowed to continue; however, it is depicted in terms of a ritualistic, punitive and generally oppressive range of social relationships[23]. The behaviour of the "uncivilized" is expressed in terms of rituals which are made up of a mixture of religions, including Christianity, in which conformity to marriage, chastity and appealing to "god" draw upon physical punishments to be maintained.

Either way, whether people are living as "civilized" or as "savages", promoting promiscuity or preserving marriage, there is an ultimate escape

---

[21] Cf. Dr. Elizabeth Rex's, "An End Word: A New Beginning", at the conclusion of *Conception: An Icon of the Beginning* by Francis Etheredge, En Route Books and Media, 2019:

http://enroutebooksandmedia.com/conception/; she gives a timeline in which it is possible to situate these novels in the context of the development of bioethics.

[22] Aldous Huxley, *Brave New World*, Harmondsworth: Penguin Books in association with Chatto and Windus, 1932, p. 40.

[23] Huxley, *Brave New World*, pp. 94-97.

from the pain of life: the choice, it seems, is between a stultifying life of ritual or suicide for "savages", or "soma", a drug-holiday from reality, for the "civilized". The social divide between a primitive society that preserved a sense of religion and marriage and family life *and* the civilized society that used promiscuity to replace marriage and factories to replace procreation and drugs to transport anyone out of every difficulty was as rigidly structured as the fences between them. It was indeed a deliberate ploy to argue the incompatibility between a stable marriage and family life and a reasonable use of technology: 'God isn't compatible with machinery and scientific medicine and universal happiness'[24]. In other words, the book is an illustration of that simple claim that either there is a "primitive" retention of basic religious beliefs and no scientific or technological development, or there is a substitution of the worship of a technologist and the eradication of all stable relationships beginning with marriage and the family. This claim completely overlooks the historical contribution of how recognizing the structure of the universe stimulates scientific investigations and how religious scientists have contributed to the advancement of modern science; and, at the same time, there is a completely unreal disregard for the inhumanity of inventions which have destroyed so much human life and wellbeing and have had such a devastating impact on the environment. In other words, the book is clearly advancing a poor understanding of Christianity as a supernatural faith in which reason is both drawn upon and developed as well as an uncritical assimilation of science and technology as if their negative human impact is simply absorbed by drugs instead of being addressed in reality.

---

[24] Huxley, *Brave New World*, p. 183.

Prologue

Mary and Bioethics

'In the compassionate Mother, sufferers of all ages have found the purest reflection of the divine compassion that is the only consolation. For, looked at in its deepest essence, all pain, all suffering is solitude, loss of love, the wrecked happiness of the rejected. Only the "com", the "with", can heal pain'[25]. In other words, there is not so much an antidote to suffering as a "being-with" which, in its own inimitable way, draws the pain into the reality of being loved; and, in that Mary's experience goes to the depths of suffering, so there is the possibility of finding a way out of our own solitude.

It does not seem so incongruous, then, to introduce Mary into a discussion on the inter-relationship between science, pain, technology, natural societies and natural religiosity, where the latter is defined not as the supernatural intervention of God and the help which has come through the Son of God and His Church, but as the various ways that man strives to make God do his will. For Mary expresses the gratuitous nature of the gifts of God: that God freely chose Mary to be the Mother of His Son and Mary, in the mystery of her freedom, chose to accept the will of God. In other words, there is a dialogue of salvation which takes the starting point of the reality of creation as a beautifully ordered gift and unfolds salvation as a supernaturally saving-gift; and, at the same time, Mary is a wholly human person who has received, superabundantly, the completing gift of grace which was lost through the sin of Adam and restored through Christ. Just, then, as the sin of Adam and Eve denatured the whole gift of human life given to them and their descendants, so the saving grace of Christ is communicated through the history of salvation, to Mary fully and then, through the gifts of His Church, to all of us. Thus Mary offers us the concrete possibility of contemplating a

---

[25] From an essay by Joseph Ratzinger, "Hail, Full of Grace", p. 77 of *Mary: The Church at the Source,* translated by Adrian Walker and containing a second essay by Hans Urs von Balthasar, San Francisco: A Communio Book, Ignatius Press, 2005

woman who is wholly redeemed and fully human.

It is not a case of advocating a kind of naturalism whereby it is a question of understanding the nature of an un-graced woman or man; rather, it is about explicitly recognizing the interdependence of grace and nature: the one bringing the good of human nature to perfection and thus helping the fulfilment of the other. In terms of human dialogue, there is a place for an investigation of the natural nature of men and women; but, in reality, it could also be a covert way of concluding to the sufficiency of human nature as it is, as it were, in all its disfigured brilliance. In the context, however, of the controversial times in which we live, it is possible that what is more helpful is a more robust recognition of God's work of redemption. In other words, then, it is *precisely because Mary magnifies the Lord* that she offers us the possibility of an objective rejoicing in the gift of woman; and, therefore, contrary to a possible first impression she offers us the possibility of transcending stereotypes, times and places. In the reality of being a woman, mother and spouse, Mary is the antidote to a falsification of woman, whether because of appearance, position or natural talents; and, therefore, she is able to contribute to an objective evaluation of the whole woman-in-relationship: one in graced body and soul. In the very nature of Mary's mystery, however, is a kind of antidote to the problem of pain in that there is a fulfilment of the hope of happiness – but a hope of happiness that is anchored in the help of God (cf. Lk. 1: 49).

In view of the sufferings which are an inescapable expression of the human heart's response to the difficulties of life, there is an integral life and love that is expressed in the mystery of Mary: a happiness that takes up and fulfils the nature of human life. If, then, the nature of human life is to be a being-in-relationship it follows that personal happiness entails the happiness of others: 'Such extension is required because man is a social animal, and his desire is not satisfied in providing for himself but he wants to be in a position to take care of others'[26]. It is, as it were, in the very unfolding of the life and vocation of Mary that we see, first in her own family and then in the family of the Church, that her personal happiness entails the happiness of others; indeed, that just as her personal happiness is expressed in the happiness of others (cf. Jn. 2: 3) so that happiness is fulfilled in the dimensions of life to the full and the definitive enjoyment of it in eternal life.

---

[26] St. Thomas Aquinas' *Commentary on Aristotle's Nicomachean Ethics*, translated by C. I. Litzinger, Notre Dame, Indiana: Dumb Ox Books, 1993, p. 38, Book 1: Lecture IX: The Nature of Happiness, Commentary of St. Thomas, 112.

General Introduction to each Chapter of the Book

In 'today's intellectual climate, only the masculine principle counts. And that means doing, achieving results, actively planning and producing the world oneself, refusing to wait for anything upon which one would thereby become dependent, relying rather, solely on one's own abilities'[1]. Thus each chapter of this book, it could be said, is an account of God acting in time according to the genesis of salvation history which, like pregnancy, has a timing which entails a positive mentality of waiting expectantly: a waiting which is not passive in any negative or derogatory sense – but a waiting which is full of the wisdom of waiting. Our times, then, are losing the wisdom of waiting and many bioethical problems are arising out of this impatience which has gone beyond being personal and become a kind of epidemic; and, indeed, it is almost as if there is a kind of maternity waiting to be recovered – not just the maternity of motherhood but the motherhood of waiting on God. These chapters are about that "slow" progress of God which is yet sure and steeped in love; and, at the same time, disregards nothing of the human creativity which God Himself has given freely to His creatures. Like these chapters, however, human creativity becomes more effective for being ennobled by recognizing that each person comes to exist and to develop in collaboration with God (cf. Gn. 4: 1; Ps. 136); and, therefore, it is not just that we need medical insight and advances but that we also need the patience and respect that becomes human interventions which are ordered to the action of God.

---

[1] Ratzinger, p. 16 of *Mary: The Church at the Source*, 2005.

Why does God not crash through human history like superman? In virtue, then, of the mystery of divine patience which acts in such a way that it is possible that no one is lost – beginning in what seems remote and unlikely places may well help us to recover a sense of hoping in God as fundamental to doing good in the crises of the day.

General Introduction to Chapter One: A New Context to Marriage. The desire to marry is both ever old and ever new. Marriage is old in terms of its reality existing from the beginning as a primordial expression of the relationship between a man and a woman: of their desire to be 'one flesh' (Gn. 2: 24); and new in that as time passes, there is the renewal of marriage in Christ, an increasingly modern appreciation of the personal nature and vocation of marriage and the ever fresh choice of the two who want to begin a life together. In the modern religious grasp of human love there is an increasing recognition of what has long been celebrated but, perhaps, poorly recognized: 'Let us be existential; let us see that the love between man and woman is a specific category and type of love, even if we prescind from the sphere of sex, that it is a beautiful and glorious reality that is destined by God's will to play a fundamental role in man's life, and that this love is the classical motive for marriage, that marriage is precisely the fulfilment of this [spousal] love'[2].

There is, however, a negative dimension to the new context for marriage in the situation of "today". On the one hand there is the increasing round-a-bout of relationships where, without planning it, people go from one relationship to another and from one marriage to another; and, in one sense, the "other" person has become the "mirage" of happiness: being happy involves the increasingly unreal expectation that it will come from finding the "right" person. On the other hand, then, losing the meaning of love

---

[2] Dietrich von Hildebrand, *The Encyclical Humanae Vitae: A Sign of Contradiction*, Steubenville: Hildebrand Press, 2018, p. 12.

restores the search for it. Thus there is, as it were, the challenge of love: What is love? What are the dimensions of human happiness? What is it in the turn and turn again to the hope of human happiness that "images" the possibility of a happiness beyond all the disappointments and imperfections of the human person?

Beginning again, then, with an enquiry about the nature of man and woman, marriage and family life is also about beginning from the beginning and yet taking a start from the Christian watershed which re-founds the language of the covenant, the promise of God to do good, in the reality of the Christian sacrament of marriage in which God is present and active: making present the paschal mystery in which Jesus Christ brings new life out of the sufferings and "deaths" of everyday life's humiliations, tragedies and trials. Bioethics is not, then, superimposed upon the reality of human life; rather, just as the human person is a bioethical word[3] so bioethics springs from the deepest account of human life, traversing the problems of meaning and suffering as indeed drawing afresh on the whole reality of an integral account of the human being-in-relationship: to God and to each other.

This first chapter takes us into the transition that brings new life to the mystery of marriage; and, as such, provides a renewable source of energy for the difficulties and sufferings of marriage and family life.

General Introduction to Chapter Two: A Marian Triptych[4]. Why call the next three chapters a "Marian Triptych"? On the one hand a triptych is normally a set of three paintings and, therefore, what presents itself is a threefold image of what is visible. What is visible almost immediately recalls the Son of God who became visible through the mystery of the *Incarnation*

---

[3] Cf. Francis Etheredge, *The Human Person: A Bioethical Word*, St. Louis: En Route Books and Media, 2017.

[4] This Marian Triptych has been published by Fr. David Meconi of the *Homiletic and Pastoral Review* (2019:
https://www.hprweb.com/author/francis-etheredge/.

(cf. Jn. 1: 14); and, therefore, what is visible is what is invisible: the body manifesting the soul: man and woman showing forth the mystery of relationship: the human manifesting the divine. In other words the reference point is not the discussion, however interesting, but the reality of personhood: both human and divine; and, in this instance, what is to be considered is that Mary is the concrete choice of God. In a particular way, then, the concrete choice of God comes with an enduring significance in terms of the constant challenge to understand the threefold providential workings of God: in the relational reality of womanhood, in the dynamic mystery of Christ and His Church and in terms of the significance of the times in which we live.

Without, however, distinguishing any one of these it is probably significant for all three that Cardinal Ratzinger, now Pope Emeritus Benedict XVI, says: 'If the misery of contemporary man is his increasing disintegration into *mere* bios and mere rationality, Marian piety could work against this "decomposition" and help man to rediscover unity in the center, from the heart'[5]. On the one hand there is the vividly painful reality of what happens when 'bios' and 'rationality' are disintegrated. In the case of "using" the human embryonic child for experiments there is a kind of dichotomy between the child and "biological life"; and, therefore, "rationality" in this case is an almost active denial of the full humanity of the subject – for human conception is the start of a process of maturing the manifestation of the person conceived. It is as if the very fact that a human child can be manipulated changes the reality of what is in the dish; but, tragically, nothing has changed in terms of the child: the conception of a human person is an irreversible reality. On the other hand, 'Marian piety could work against this "decomposition" and help man to rediscover unity in the center, from the heart.' In front of the mystery of human conception, a mystery involving the action of God, a response from the welcoming depths of the woman's being

---

[5] Joseph Ratzinger, p. 36 of *Mary: The Church at the Source*, 2005.

is expressed in the words of Eve: 'I have gotten a man with the help of God' (Gn. 4: 1); and, even if the woman participates in the fall of creation as well as the man, so the experience of these sufferings is not without significance for the healing of the wounds of sin (cf. Gn. 3: 16-19). Just as Mary is chosen as the Mother of the Lord, in view of the gift of spousal womanhood, so women are entrusted with the welcome of a child; indeed, it may well lie in the psychological makeup of the woman to grasp the wholeness of human being as a gift to be welcomed and nurtured.

In a world, then, which "uses" the emotional argument to exploit people's sympathy in response to suffering and to advance an answer of death to life's difficulties, whether it be abortion, euthanasia or destructive embryo experimentation, Mary's experience of suffering as a participation in the suffering of Christ is also an indication of where we are to find the roots of an answer to the black holes which threaten to destroy us. At the same time, however, as there is a medicine of life which seeks to cure and to alleviate the suffering we experience there is, too, a resurrection of meaning that awaits those who persevere in praying through the problems of life.

General Introduction to Chapter Three: Hope. 'It is on the path shown by this ... [sign of the woman] that we follow the trail of hope toward Christ, who guides the ways of history through this sign that points the way'[6]; and, more specifically, 'the originality of Mary's role of mediation consists in its maternal character, which aligns it with Christ's being born ever anew in the world'[7]. In other words, in a world in which motherhood is disfigured the motherhood of Mary is a sign of hope in the ever anew coming of Christ. But what is this hope?

As St. Paul says, 'faith, hope, love abide, these three; but the greatest of these is love' (1 Cor. 13: 13). Faith, hope and love are 'infused by God ... to

---

[6] Joseph Ratzinger, p. 53 of *Mary: The Church at the Source,* 2005.

[7] Joseph Ratzinger, p. 55 of *Mary: The Church at the Source,* 2005.

make [the faithful] ... capable of acting as his children and of meriting eternal life' (CCC, 1813). By faith 'we believe in God and believe all that he has said and revealed to us, and that Holy Church proposes for our belief, because he is truth itself' (CCC, 1814). By hope 'we desire the kingdom of heaven and eternal life as our happiness, placing our truth in Christ's promises and relying not on our own strength, but on the help of the grace of the Holy Spirit' (CCC, 1817). By love or charity 'we love God above all things for his own sake, and our neighbour as ourselves for the love of God' (CCC, 1822).

The axis of eternal life, like the axes of human development, is to make possible the daily living in which all our joys and sorrows are experienced. Thus there is no contradiction between daily life and its demands and how 'the fruits of our nature and our enterprise ... [are] cleansed ... from the stain of sin, illuminated and transfigured, when Christ presents to his Father an eternal and universal kingdom' of 'justice, love and peace' (*Gaudium et Spes*, 39). In other words there is every reason to pursue the goal of alleviating, if possible, those suffering from infertility, debilitating diseases or the alleviation of the experience of the sufferings that can turn people towards the possibility of suicide. Respect for life, however, challenges us to recognize that there is an ethical framework within which we work and which guides, practically, the development of bioethics; and, therefore, research into infertility is guided by seeking a way of helping which expresses the reality of the unitive and procreative significance of spousal love (*Humanae Vitae*, 12) and the sacredness of human life (*Humanae Vitae*, 13).

In terms of the specific theme of bioethics, then, we need to hope in the help of God in terms of the very medical procedures which are in accordance with His will; as, indeed, there is no certainty as to whether or not an illness can be cured, infertility remedied or suffering illuminated and alleviated. In other words there is a possibility of passing through darkness, at times, in the course of persevering with our difficulties; indeed, 'For Mary, as for Abraham, faith is trust in, and obedience to, God, even when he leads her through darkness'[8]. Even in the case of persevering in the pursuit of an ethically upright way of helping those who have been frozen, as human embryos, there is a kind of "ethical darkness" until it becomes clear that a way can or cannot be found to help them; and, indeed, what are we to say of those who have been frozen in this unspeakably tragic disregard for the ongoing unfolding of their lives? Prayer and perseverance, then, are essential in the hope-led search to help those who suffer; and, indeed, like the wise men of the east, we need to search every aspect of human wisdom as well as appeal, constantly, to God to guide us.

General Introduction to Chapter Four: Prayer. 'The Fathers of the Church say that prayer, properly understood, is nothing other than becoming a longing for God. In Mary this petition has been granted: she is, as it were, the open vessel of longing, in which life becomes prayer and prayer becomes life'[9]. 'We can ask ourselves: when I pray, do I open myself to the cry of so many close and distant persons? Or do I think of prayer as a sort of anaesthesia, to be able to be more tranquil?'[10]

Thinking of prayer, then, it can seem that it is about "moments of prayer" as if, owing to the crises of life, we turn to God in prayer because it is impossible that anyone else can understand or help us; and, indeed, this remains a consolation in the depths of human suffering: that the aloneness we experience is a kind of *aloneness with the alone*[11]. I particularly remember

---

[8] Joseph Ratzinger, p. 49 of *Mary: The Church at the Source*, 2005.

[9] Joseph Ratzinger, p. 15 of *Mary: The Church at the Source*, 2005.

[10] Pope Francis, "Pope at General Audience on Jesus' way to Pray (Full Text)", Virginia Forrester, February 13th: https://zenit.org/articles/pope-at-general-audience-on-jesus-way-to-pray-full-text/.

[11] Cf. A song composed by Kiko Arguello of the Neocatechumenal Way is called 'Sola a Solo' (Amsterdam, 30th April, 2005): 'Sola a Solo, under the cross, Mary, who can separate you? Virgin alone …'. Cf. also Francis Etheredge, *The Prayerful Kiss*, from enroutebooksandmedia, 2019: enroutebooksandmedia.com/theprayerfulkiss/.

an intensely difficult time in my married life when the meeting in which this song was sung and shared gave me an opportunity to talk about the pain that existed within; and, almost on the basis of solely being listened to, the experience of this pain being heard seemed to lance and help it. There are depths to the human heart, then, that we almost cannot reach in the human terms of analysing and explaining; but, in the experience of "being with", there is a kind of transcendence of the isolating, almost socially suffocating sense of being unheard by others.

In the intense experiences that people go through in the prolonged experience of pain, whether it is the pain of infertility, of the suffering inflicted or otherwise brought about by others, the length and depth of an illness, there is a kind of point of darkness that is almost out of reach and incapable of being communicated; and yet, this interior pain is not unintelligible and belongs to the profoundly shared human experiences which are, in a way, the current that runs through human nature and throughout the whole "liturgically real" experience of the *Paschal death and Resurrection of Jesus Christ*. In other words, although it seems to us at different times that there are experiences which are "unshareable", the nature of the suffering between Christ and His Mother exists as a kind of relationship within which it is possible for all human experience to find a home. In the difficult if not impossible experiences which often constitute the human kernel of a "bioethical" crisis, dilemma or apparent dead end, there is a prayer which rises out of the depths of the human spirit which is touched by the Spirit of God: a prayer which opens and engages in a wordless dialogue with God in a way that makes help possible.

General Introduction to Chapter Five: Gratitude and Ingratitude. There is, as it were, a response of gratitude to human existence: a sense of gratitude which recognizes that being in existence is a gift, whole and entire;

indeed, in the words of Hans Urs von Balthasar, gratitude is twofold: 'It is essential, however, to bear in mind that the two trajectories of thanksgiving – towards one's parents and toward God – are not in competition; they exist next to, together with, and in each other'[12]. This gratitude, moreover, is essentially human: 'Therefore, insofar as he is "born of woman", Christ owes thanks for himself to his Mother, and he has to do so in order to be man in the full sense'[13]. In 'Mary, the Church is embodied even before being organized in Peter'; and, therefore, 'The Church is primarily feminine because her primary, all-encompassing truth is her ontological gratitude, which both receives the gift [of salvation] and passes it on'[14]. In other words we are all caught up in a movement of gratitude that begins in being given existence as a gift and grows as it is permeated by the grace of salvation which comes through Christ and His Church.

By contrast, then, there is an attitude which expresses a desire 'to owe no one thanks for himself' and, indeed, to react to man's 'vexing dependence as an alienating bondage' to the extent of claiming that 'man is more himself the less he gets his being from others'[15]. Thus there is the possibility that the deepest part of the pain of being imperfectly human is, as it were, an ingratitude which expresses the rejection of the *being-gift* of human personhood. It therefore seems possible that the problem that manifests itself in a variety of ways, particularly in the bioethical crises that confront us, is the problem of an existential ingratitude: an ingratitude that expresses a fundamental revolt against the gift of being; however, rejecting the gift of being does not remedy the problem of being who I am but, rather, constitutes an almost insurmountable obstacle to the reconciliation that brings peace: the reconciliation between the person and the reception, as it were, of human

---

[12] Hans Urs von Balthasar, "The Marian Mold of the Church", p. 127 of *Mary: The Church at the Source*, 2005.

[13] *Ibid*, p. 130.

[14] *Ibid*, p. 140.

[15] *Ibid*, p. 126.

existence from God and from our parents.

Drawing on these initial thoughts, then, it can be said that a discussion on the dogma of the *Immaculate Conception* is a discussion which both ontologically roots Mary's response of gratitude to God and, at the same time, reveals essential elements to a modern discussion of bioethical questions. In other words, both because Mary is a true human being, created as we are, and because she has been the subject of a dogmatic pronouncement about her original sinlessness, she is the perfect exponent, as it were, of our own beginning. There is the view, however, that 'The phrase, [preserved free from all stain of original sin] "in the first instant of her Conception," avoided debates which belong more properly to biology'[16]. Nevertheless, as it was 'The person of Mary, not merely her soul, [that] was the subject of the immunity from original sin'[17], it follows that it is possible to explore this dogma in the hope of it indirectly illuminating the question of human conception of the first instant of Mary's conception; for the unity of the human person, one in body and soul (*Gaudium et Spes*, 14), implies the possibility that the moment of Mary's conception being "uniquely required" by the 'singular grace and privilege granted by Almighty God, in view of the merits of Jesus Christ, the Saviour of the human race'[18] is the first instant of fertilization. In view, however, of the possibility that Pope Pius IX did not have the first instant of conception in mind when he promulgated *Ineffabilis Deus*, he says that his predecessors

> 'Definitely and clearly ... taught that the feast [of the *Immaculate Conception*] was held in honor of the conception of the Virgin'[19].

---

[16] Adapted quotation from "Immaculate Conception, The", p. 179 of Michael O'Carroll's, *Theotokos: A Theological Encyclopedia of the Blessed Virgin Mary*, Eugene, Oregon: Wipf and Stock Publishers, 2000.

[17] O'Carroll, "Immaculate Conception", p. 179.

[18] O'Carroll quoting from 'The Papal Bull, *Ineffabilis Deus*', p. 179.

[19] *Ineffabilis Deus*: http://www.newadvent.org/library/docs_pi09id.htm.

General Introduction to Chapter Six: Male and Female. There is a concreteness to the action of God. 'If the concrete were not essential to Christianity, if the Lord were not truly flesh but "merely the product of an idea," then Mary might have been only spiritually pregnant, Adrienne reasons; but the reality of the incarnation requires that she "feel his weight in her body and, after His birth, in her arms"'[20]. Indeed, Hans Urs von Balthasar makes this very clear himself when he says that in the *Incarnation* 'we are of course dealing with a genuine physical fecundation'[21] in which 'the Spirit - faithful to the task received from the triune Father – is primarily an envoy, the obedient bearer of the divine seed and of its actualization in the womb of Mary's Yes'[22]. In other words the word of God is visibly effective in bringing to exist what did not exist.

Just as in the beginning the word of God brought creation to exist so the *Incarnation of the Son of God* shows forth the fruitful action of the Holy Spirit. While, then, this generation may argue that man, male and female, is a provisional project which may be adapted at will, what is actually astonishing is the unity-in-diversity of precisely those human relationships which originate with the beginning of man's being, male or female. Indeed, just like the triangular based tetrahedron which is a basic building block of the universe, the triangular relationship of God, man and woman is amazingly capable of building up a marvellously complex society from the simplest complementarity, as it were, of man and woman: a complementarity which reaches from the depths of eternity into the foundations of the beginning and is again taken up in the wondrous work of salvation drawing

---

[20] Michele M. Schumacher, "A Speyrian Theology of the Body", quoting from Adrienne von Speyr, *Maria in der Erlosung*, 35, p. 264 of pp. 255-292 in the book *The Virgin Mary and Theology of the Body*, edited by Donald H. Calloway, Pennsylvania: Ascension Press, 2007.

[21] *Explorations in Theology: V: Man is Created*, translated by Adrian Walker, San Francisco: Ignatius Press, 2014, p. 177.

[22] *Explorations in Theology: V: Man is Created*, p. 179.

on the virginal manhood of Jesus Christ and the spousal mystery of the Virgin Mary. When, according to St. John Paul II, God pondered the divine 'We' - man and woman came forth:

> 'Before creating man, the Creator withdraws as it were into himself, in order to seek the pattern and inspiration in the mystery of his Being, which is already here disclosed as the divine "We". From this mystery the human being comes forth by an act of creation: *"God created man in his own image,* in the image of God he created him; male and female he created them" (*Gen* 1:27)' (*Letter to Families*)[23].

The dialogue of the sexes is, therefore, a part of the very revelation of God: a revelation which, in a sense, involves the unity-in-diversity of human relationships; indeed, that it is necessary to ponder the mystery of man and woman anew: anew because of the multitude of ways that this "Icon" of the Blessed Trinity is being disfigured and needs to be reclaimed to be the healing anointing it is.

General Introduction to Chapter Seven: The Challenge to Believe God is for us! Sufferings are so brutal that they dismantle us and tempt us to believe that God does not exist or does not even know about us - never mind loves us! Or that if there is a God who witnesses the misery of our lives and does nothing – How terrible if that were true! Sufferings bring us to death: to want to die and to be done with a life so unbearably unacceptable to us! Sufferings are a language unintelligible to us: Does not every ounce of human goodness say – "If I were God would I let human freedom be as free as God allows?!"

At the same time, in terms of the difficulties encountered in the culture

---

[23] The English title of this letter is given in the text as the Latin title is less well known, which is *Gratissimam Sane*.

in which we live, there are three problems which imply a breakdown in the possibility of understanding the value and mystery of marriage: the availability and use of pornography, suggesting a blindness to the existence of the person as an 'incarnate spirit' (*Familiaris Consortio*, 11); the temporary 'hookup' culture which seems to substitute sex for conversational intimacy; and, thirdly, the cycle of divorce and its impact on the possibility of believing in marriage as being husband and wife, life-long and open to life[24].

Add to this the mentality that everything and everyone is "plastic", that every problem requires a technological intervention and that there are no principles, only arguments that more or less appeal to human sympathy; and, therefore, *in vitro* fertilization, appealing to the plight of infertile couples, does not actually offer a remedy for infertility but, rather, seizes the opportunity to intervene, technologically, as if the problem does not involve the whole human being: the whole human being of the man, woman or child.

Thus there is nothing neutral about the "situation" of today and its many facets: facets that are actually profoundly implicated in the deepest questions of the human heart, the experience of agonizing sufferings and the problem of being unable to reason to and from the reality of the whole human being-in-relationship. The response of love, truth and goodness to all the difficulties of life has to begin again, then, in the new context of the regeneration of human life and culture, drawing on all that is good, true and loving in the practices of "today"; but, at the same time, this new beginning is not wholly new but is integral to the new beginning which God is constantly bringing about and which we are continually rediscovering. In the literature drawn from the Early Fathers of the Church it is possible to come upon unexpected thoughts in the writing of the Church Fathers and, as we increasingly recognize, the great Mother of the Church. Thus, in an excerpt

---

[24] Randall Woodard, "Our Current Youth Culture and its Upcoming Impact on Successful Marriages", https://www.hprweb.com/2017/11/our-current-youth-culture-and-its-upcoming-impact-on-successful-marriages/.

from the work of St. Cyril of Alexandria, (d. 444), we read that 'Many notable things were accomplished in this one sign [when Christ turned water into wine at the marriage feast of Cana], his first sign. Honorable marriage is sanctified, and the curse pronounced against woman is overcome. Women will no longer bear children in sorrow, since Christ has blessed the very beginning of our lives'[25]. Thoughts that St. John Henry Newman continues, both recognizing how 'polygamy and divorce' were 'detrimental to the dignity of women'[26] and asserting that 'women "shall be saved through the Child-bearing, that is, through the birth of Christ from Mary, which was a blessing, as on all mankind, so peculiarly upon the woman"'[27].

---

[25] Footnote 7: *Commentary on John* 2, 1; PG 73, 228 cited on p. 245 of *Mary and the Fathers of the Church: The Blessed Virgin Mary in Patristic Thought*, by Luigi Gambero, translated by Thomas Buffer, San Francisco: Ignatius Press, 1999.

[26] From the commentary by Philip Boyce, p. 71 of *Mary: The Virgin Mary in the Life and Writings of John Henry Newman*.

[27] Quoted from *Parochial and Plain Sermons*, Vo. II, p. 131 in *Mary: The Virgin Mary in the Life and Writings of John Henry Newman*, p. 71.

FOREWORD TO CHAPTER ONE: THE SANCTITY OF MARRIAGE

Dr. Mary Anne Urlakis

Love one another as I have loved you (John 15:12).

Introduction: A Profound Mystery

The sacramental nature of marriage is a profound mystery, one which transcends space, time, culture, and even eternity. In this initial chapter of the text <u>Mary and Bioethics</u>, Francis Etheredge explores the ancient antecedents of Christian marriage, discussing both its Jewish roots and its early Christian transformation. Throughout this essay, Etheredge carefully balances reflection on the historical, philosophical, and theological roots of marriage, evident in Sacred Scripture and historical documents, with a consideration for the contemporary reality of Christian marriage in the Twenty-first Century. Exploring the mystical and dynamic reality of the marriage bond between the Blessed Virgin Mary and St. Joseph, Etheredge embarks upon a path which exemplifies the concept that, "The Christian sacrament of marriage is 'consonant' with that of Judaism, and at the same time, a renewal and development of it."[1] Like a finely braided cord, the three strands of Sacred Scripture, Sacred Tradition, and authentic Magisterial teaching are woven through this essay, supporting and binding it together.

---

[1] Etheredge, *Mary and Bioethics*, Chapter 1.

From the beginning, papal teaching regarding the sacramental nature of marriage has emphasized this renewal and development of marriage as a purposeful and divinely instituted action. In *Arcanum*, the first papal encyclical devoted to the subject of contemporary marriage, beginning in the nineteenth century, Pope Leo XIII reminds us:

> "To the Apostles, indeed, as our masters, are to be referred the doctrines which "our holy Fathers, the Councils, and the Tradition of the Universal Church have always taught," namely, that Christ our Lord raised marriage to the dignity of a sacrament; that to husband and wife, guarded and strengthened by the heavenly grace which His merits gained for them, He gave power to attain holiness in the married state; and that, in a wondrous way, making marriage an example of the mystical union between Himself and His Church, He not only perfected that love which is according to nature, but also made the naturally indivisible union of one man with one woman far more perfect through the bond of heavenly love."[2]

Ancient history demonstrates clearly that while divine in origin, the institution of marriage has suffered over the centuries from familiar assaults, often the same secular attacks levied against the nuptial bond today. It is through Christ and His obedience to the plan of the Heavenly Father that the bond of matrimony was elevated as a sacrament and thus instituted as a gift for the sanctification of humanity; a bequest which perpetually reflects Christ's own indissoluble bond with His Bride, the Church, the life-giving bond forged for all eternity through His Cross and Resurrection.

---

[2] Pope Leo XIII, *Arcanum*, [Section 9] promulgated February 10, 1880. *Arcanum* has long been hailed as the papal document which is the foundational precursor to both Pope Pius XI's 1930's encyclical *Casti connubii* and Pope St. Paul VI's 1968 *Humanae vitae*.

In Section 8 of *Arcanum*, Pope Leo XIII addresses this consideration:

"So manifold being the vices and so great the ignominies with which marriage was defiled, an alleviation and a remedy were at length bestowed from on high. Jesus Christ, who restored our human dignity and who perfected the Mosaic law, applied early in His ministry no little solicitude to the question of marriage. He ennobled the marriage in Cana of Galilee by His presence, and made it memorable by the first of the miracles which he wrought; and for this reason, even from that day forth, it seemed as if the beginning of a new holiness had been conferred on human marriages. Later on, He brought back matrimony to the nobility of its primeval origin...."[3]

At the heart of this chapter is Etheredge's proposition, "that the sacraments, *taking up, as it were, a prior Jewish form*, are revealed in a way that almost indicates a sensitivity to the very "passage" of the transformation of Judaism into Christianity."[4] Etheredge's thesis is that marriage-specifically the marriage between the Blessed Virgin and St. Joseph stands at the beginning of this transformation. In turn, the transformation of marriage parallels the movement from Passover lamb of the Exodus in the Old Testament to the ultimate, eternally redemptive, Paschal Sacrifice by the *Incarnate Word* in Jesus Christ's Crucifixion, death, and Resurrection, perpetually ever-present in the Sacrament of Holy Eucharist.

Thus, when we consider the topic of authentic marriage it is essential to reflect on its sacred nature. For whether we are speaking of the marriage between the Blessed Virgin and St. Joseph, or the marriage of Tobias and Sarah, of Saints Anna and Joachim, Zachariah and Elizabeth, Louis and Zeli Martin, or the sacramental union of our parents, grandparents and relatives,

---

[3] Pope Leo III, *Arcanum*, Section 8.

[4] Etheredge, *Mary and Bioethics*, Chapter 1.

a hallmark of authentic marital union is its sacred character. That sacred nature is a necessary attribute, which is both gift from the Triune God Who ordained and instituted the bond of marriage, and simultaneously a reflection back and a participation in the Divine Mystery which is the Espousal of Christ and His Bride the Church.

The Sixteenth-Century Carmelite mystic and Doctor of the Church, St. John of the Cross, eloquently captures the beauty of this reality in his poem *The Romances*. The Third Stanza, entitled "On Creation" expresses a dialogue between the Father and the Son, and begins:

> "My Son, I wish to give you
> a bride who will love you.
> Because of you she will deserve
> to share our company,
> and eat at our table,
> the same bread that I eat,
> that she may know the good
> I have in such a Son;
> And rejoice with me
> In your grace and fullness.
> "I am very grateful,"
> The Son answered;
> "I will show my brightness
> to the bride you give me,
> so that by it she may see
> how great my Father is,
> and how I have received
> my being from your being.
> I will hold her in my arms,
> And she will burn with your love,
> And with external delight

> She will exalt your goodness."[5]

Sacramental Marriage is a unity, a covenant which is sacred in nature. As such it derives its authority from the One Who is its origin and author. In this essay, Etheredge explores the "perspective of the cross" present within each authentic marriage, for it is only through the cross that the couple joined in matrimony can live the depth of their vocation on earth and come to the fullness of grace that will be revealed in eternity. As Etheredge demonstrates throughout this essay, the language of marriage is woven as imagery throughout Sacred Scripture, thus continually highlighting the transcendent and everlasting bond that is an implicit component of a relationship forged via the covenant.

Chapter 62 of the Book of Isaiah exemplifies this language and the covenantal relationship it signifies: "As a young man marries a virgin, your Builder shall marry you; and as a bridegroom rejoices in his bride so shall your God rejoice in you (Isaiah 62:5)."[6] The covenantal relationship of marriage is thus one of joy – a unity which transforms that which was once called "Desolate" and "Forsaken" into a new reality that is the Lord's "Delight," His Beloved "Espoused."[7] The covenantal relationship between God and humankind is eternally manifest, present uniquely in both the relationship between Christ the Bridegroom and His Bride the Church, and equally present and manifest in the authentic sacramental marital union between Christian spouses. In both the nuptial bond between Christ and the Church and the marital bond between a human man and woman, the sacramental nature of the covenant is manifest by sacramental realities: the waters of baptism and the Mystery of the Cross.

---

[5] St. John of the Cross, The Romances, in <u>The Collected Works of St. John of the Cross</u>, Translated by Kieran Kavanaugh and Otilio Rodriguez, ICS Publications, 1991, p. 62.

[6] Isaiah 62:5, New American Standard Bible translation.

[7] Isaiah 62:4, New American Standard Bible translation.

Pope St. John Paul elucidates the intrinsic covenantal relationship of sacramental marriage in his Section 13 of his 1981 Apostolic Exhortation, *Familiaris Consortio*, stating:

> "Indeed, by means of baptism, man and woman are definitively placed within the new and eternal covenant, in the spousal covenant of Christ with the Church. And it is because of this indestructible insertion that the intimate community of conjugal life and love, founded by the Creator, is elevated and assumed into the spousal charity of Christ, sustained and enriched by his redeeming power."[8]

Just as at the Wedding at Cana, Christ changes the water contained in six stone ceremonial jars into wine, and in so doing prefigures the Holy Eucharist and its real participation in His salvific suffering, death, and Resurrection, so too each couple received at baptism the foundational grace requisite to enter into a sacramental marriage.

Pope St. John Paul continues in *Familiaris Consortio*:

> "By virtue of the sacramentality of their marriage, spouses are bound to one another in the most profoundly indissoluble manner. Their belonging to each other is the real representation, by means of the sacramental sign, of the very relationship of Christ and his Church."[9]

While it is through the waters of Baptism that the spouses enter into the eternal covenant with the Blessed Trinity, it is through the cross, and only through the cross that the covenant of Christian Marriage is ratified. "Spouses are therefore the permanent reminder to the Church of what

---

[8] Pope St. John Paul II, *Familiaris Consortio*, Apostolic Exhortation, November 22, 1981, Section 13, p. 29.

[9] *Ibid.*

happened on the cross, they are for one another and for the children witnesses to the salvation in which the sacrament makes them sharers."[10]

The New and Eternal Covenant between God and humankind was ratified once and for all on the Cross, sealed for all Eternity in the Precious Blood of the Unblemished Lamb- Christ the Bridegroom. As such, each human sacramental marriage also shares uniquely in this act of ratification. In his 1994 *Letter to Families*, entitled *Gratissimam Sane*, Pope St. John Paul refers to this great truth:

> "Jesus appeals to "the beginning," seeing at the very origins of creation God's plan, on which the family is based, and through the family, the entire history of humanity. What marriage is in nature becomes, by the will of Christ, as true sacrament of the new covenant, sealed by the blood of Christ the Redeemer. *Spouses and families, remember at what price you have been "bought!"* (cf. 1 Cor 6:20)."[11]

Symbols of the Sacramental Reality of Marriage

The transcendental reality of this covenantal relationship is reflected in various symbols associated with the Rite of Christian Marriage in both the Eastern and Western Traditions of the Church. The exchange of rings is among the oldest and most ubiquitous of traditions regarding marriage, with historical testimony verifying the practice among the diversity of the world's cultures including textual references reflected in six-thousand-year-old Egyptian papyrus manuscripts. Evidence of the exchange of wedding rings has been likewise documented in ancient Jewish and Christian cultures, as well as those of ancient Rome and Greece. Over the millennia, the wedding

---

[10] *Ibid.*

[11] Pope St. John Paul II, *Gratissimam Sane*, February 2, 1994, Chapter 2, Section 18, p. 68.

ring has symbolized both the eternal and perpetual obligation of the marital commitment as well as the fidelity which – even in pagan unions – was idealized as a necessary force- binding the love of the spouses together.

As is the case with other symbols that were essentially transformed by incorporation into Christian sacramental life, the wedding ring also gained a deeper significance within the Rite of Christian Marriage. While retaining its original symbolic connotations of representing the eternal and faithful love of the spouses, it also came to reflect the eternal and faithful love of God- reflecting the Divine Origin of Love Itself, and the eternality of the covenant now ratified. Similarly, the hole in the center of the Christian wedding ring is no longer perceived as empty space, but rather a portal to eternity, as

---

[12] Photograph, August 31, 2019, courtesy of Mary Anne Urlakis: The window is in Saints Peter & Paul Catholic Church, Door County, Wisconsin.

Christian spouses are charged with the responsibility of aiding and accompanying each other along the way to Eternal Life in Christ.

Sacramental grace is gift which allows Christian couples to live the mystery of marriage day after day, year after year, embracing each other in the mundane trials of the moment, while focusing on the eternal goal of salvation in Christ. The Croatians have an ancient tradition which beautifully reflects this reality. During the Nuptial Mass, at the exchange of vows, the priest stands between the betrothed holding a crucifix. The couple join hands, weaving them around the crucifix, symbolically attesting – in both word and deed – to their life-long commitment to building a marriage founded upon a relationship with Jesus Christ, through the power of the cross. Similarly, in the Eastern Christian tradition, the couples place their hands – under the priest's stole – on the Book of the Gospels, demonstrating visibly that their marriage is to be founded on the Living Word of God.

Throughout Christianity, there is recognition that authentic Sacramental Marriage is a life-long journey through the mystery of love- a sacrificial love that necessarily demands a commitment to daily martyrdom. Both Eastern and Western Christian Art reflect the centrality of the symbolic crown of martyrdom to sacramental married life. In the Eastern Church's sacramental nuptial celebration, the Rite of Crowning of the couple is canonically essential to the ritual of marriage- in fact, in the Byzantine Church, the Sacrament of Matrimony is called the "Mystery of Crowning." The prayers and hymns all celebrate the sacrificial nature of married love and call upon the glorious martyrs in Heaven to intercede for the newly married couple.

George and Lorraine

From the Theoretical to the Concrete

I have found the One Whom my Soul Loves (Song of Solomon 3:4).

Sacrificial in character, authentic Christian Marriage is ultimately covenantal- receiving both its origin and achieving its destiny through its relationship to God and His Eternal Covenant with humanity. The *Catechism of the Catholic Church* teaches:

> "The consent by which the spouses mutually give and receive one another is sealed by God himself. From their covenant arises "an institution, confirmed by the divine law, . . . even in the eyes of society." The covenant between the spouses is integrated into God's covenant with man: "Authentic married love is caught up into divine love."[14]

---

[13] Photograph courtesy of Mary Anne Urlakis.
[14] Catechism of the Catholic Church, 1639.

Theologically, philosophically, and theoretically, it is important to examine the nature of sacramental marriage; yet it is in the concrete that each of these salient aspects are exemplified. I often ponder in awe of the sacramental grace that floods my own marriage with joy, mercy, and love. Thirty years of sacramental wedlock have given me a profound reverence for the mystery of Divine Grace which binds two imperfect souls in a life-long commitment to building a domestic church for the sanctification of not only each other, but also for the progeny with whom they are gifted. Yet, it is when pondering the sacrificial nature of my parents' marital bond and the union of their own parents that I concretely witness both the sacrificial love of spouses and the mystery of Divine Grace acutely manifest.

The generation of my grandparents

George and Lorraine, my parents, were part of the "Greatest Generation," - a generation known for resilience, faith, and industrious ingenuity. Dad and Mom were born during the Great Depression, and like so many others of their era they both experienced tragedy, profound loss, and separation from loved ones within the first five years of their respective childhoods. Lorraine's father died when she was barely three; a death which fell within a decade of the deaths of her three eldest brothers and within the year of the birth and death of her baby sister. I grew up hearing stories of my grandmother, Angeline's, struggle to care for her family as widow during the Great Depression. By day, Angeline ran a small Polish grocery market; by night, she scrubbed floors in Chicago's Union Train Station. Staunchly Catholic, Angeline raised my mom and her sister to know, love, and serve God, understanding that the cross was a necessary component of the mystery of life.

My dad spent his first nine months of life in a Chicago orphanage- though as an adult the revelation of his adoption did not come until he was in his sixties. Years after my father's death, a relative shared what she knew of my father, George's adoption-history that my dad only learned on the other side of eternity. My grandfather, Hartford, was also adopted, and when he and his beloved spouse lost a baby girl during a traumatic delivery, their hearts were broken. Hartford and Gudrun had been praying for a child for a long time and after the death of their precious little one, they were devastated to hear that Gudrun would never again be able to conceive and deliver a child. Months passed, but the pain and longing for a child remained. Like Hartford's own adoptive parents, he and Gudrun decided that the fruit of their love was meant to embrace an orphaned child. Gudrun and Hartford visited the nearby orphanage, seeking a newborn baby girl. What they found instead, was my father, standing in a crib with arms outstretched, beckoning to them. Their hearts melted. From that day on, in 1929, their forever-family was forged in love and commitment.

Sixteen years later, my dad joined the U.S. Navy in his junior year of high school, and after receiving basic training through the Great Lakes Naval Base in Chicago, Illinois, served his country as a Seaman First Class on an LST, tank landing ship[15], during the final years of WWII. Growing up, Dad would often relate stories of the train ride home for a rare weekend to visit his folks, and just what their love and support meant to him. At the tender age of eighteen, Hartford had fought as a doughboy[16] in World War I and lost one of his two brothers in France during the war; thus he was fully cognizant of the dangers to body, mind, and soul that young men face during war. He and Gudrun must have agonized and prayed as they sent their only son off to enlist in the Navy at the age of seventeen. In the post-war chaos and jubilation, like many other military men returning home, Dad sought the stability of a wife, family, and gainful employment.

---

[15] Literally, LST stands for Landing Ship, Tank.

[16] A nickname for American infantrymen.

The marriage of my parents

George and Lorraine met in Kalamazoo, Michigan in August of 1952. They were introduced to each other by my maternal aunt and uncle with whom my mother and grandmother lived. My aunt and uncle worked at the same Woolworth's Department Store which employed my father. George worked in various capacities in the store, soda fountain, stockman, clerk, as did my aunt and uncle. My mother worked as a switchboard operator for American Telephone and Telegraph- she was one of the young ladies who wore roller-skates on the job, connecting callers from one circuit to another. Surviving the Great War tended to instil a reverence for the fragility of life, and my mom and dad were acutely aware of the rare and awesome gift of true love. They met in August, were engaged in January of the following year, and in June were married at Mary Queen of Heaven Church in Chicago. At the time of the wedding, my mother was an orphan- not only had her father died when she was a toddler, but my grandmother died two months after my parents' engagement. Like many who grew up during that era, my parents were humble folk of deep faith and traditional virtue. Their married life was marked with simplicity, struggle, and joy. God granted them the blessing of three children and later eleven grandchildren.

Like her older sister, Lorraine was to be afflicted with early-onset Alzheimer's Disease, a disease process which began while I was still in high-school. Through memory loss and personality changes, the bond of love forged in the grace of sacramental marriage remained vividly vital. Shortly after my husband and I were married, we received an alarming phone call from an emergency room nurse two hours away. My parents had been traveling to visit, as we were in the middle of helping them find housing closer to us so that we could be of more assistance in their daily lives. That morning they had been in a serious head-on car crash and were severely injured. Frantically, my husband and I drove to the small-town emergency room where they had been admitted. As we walked into the emergency room, there they were in the two-bed trauma room- holding hands under the curtain that separated the gurneys on which they were laying. I spent the night in the hospital with them. In addition to his already numerous health issues, Dad had four broken ribs; Mom's trachea had hit the dashboard and her face was lacerated from its impact on the windshield. They had not been wearing seat belts, and thus were fortunate to be alive - and they knew it. We prayed through the night, rosary after rosary. They doted on each other, laughed and cried, and prayed for the poor pregnant woman who had run a stoplight and hit them. I witnessed again the profound love that Christ had poured into their marital union, a love that was ever-bound to the cross.

Love in the years of decline

As Alzheimer's inexorably claimed Lorraine's mind and personality, George's physical health also declined with cardiac disease, diabetes, and cancer. Through the years of sickness and physical death that followed, I continued to marvel at their commitment to each other and to Christ, and began to appreciate the profound presence of the love that only the supernatural grace abundant in sacramental marriage can yield. Even though George's declining health necessitated Lorraine's move to a skilled Alzheimer's unit, he continued to visit her every day, bringing an ice- cream treat, which he would feed her with tender love and care.

Shortly after her transfer, George began a five-month battle with immunoblastic multiple myeloma - an aggressive bone cancer. A close-knit family, we had been blessed to see my parents nearly daily, and George's cancer diagnosis necessitated that he immediately resides with my husband, I, and our three small children. The kids and I would drive Dad to his appointments for daily radiation treatments and monthly chemotherapy and invariably during the drive he would remind us that we needed to stop and visit "Grandma Lorraine." During one such visit, as my children - then ages four, two, and ten-months, were wreaking havoc near the nurses' station, I took a moment to step back and watch the tender exchange between my gravely-ill father and my non-communicative mother. As Dad sat there holding Mom's hand, and feeding her ice-cream, a light passed between them - a joy that was palpable. Lorraine had not spoken a word in nearly two years, yet between the two them there was a grace and a communication that transcended words - a beauty that transcended mortality, time and space. As tears welled-up in my eyes, Dad turned to me, and as if reading the questions in my heart, Dad simply said, "Isn't she beautiful? See, I still see her as she was at twenty-one." That is authentic Christian love - a marital bond blessed by God, ratified via the cross, and perfected in the Resurrection to life eternal!

As you embark upon Etheredge's essay, Chapter One: "The Holy Family: Celibacy and Marriage: A Reflection on the "Passage" from the Jewish Rite of Marriage to the Christian Sacrament of Marriage," I invite you to reflect on the sacramental marriages in your own experience - including those of parents, grandparents, and friends- blessed unions that elucidate the theoretical and practical realities presented in this salient chapter.

There are three things that last: Faith, Hope, and Love, and the greatest of these is Love, (1 Corinthians 13:13).

Mary Anne Urlakis, M.A., Ph.D, 25 April, 2020.

CHAPTER ONE:

THE HOLY FAMILY, CELIBACY AND MARRIAGE: A REFLECTION ON THE "PASSAGE" FROM THE JEWISH RITE OF MARRIAGE TO THE CHRISTIAN SACRAMENT OF MARRIAGE[1]

General Introduction to Chapter One: A New Context to Marriage. The desire to marry is both ever old and ever new. Marriage is old in terms of its reality existing from the beginning as a primordial expression of the relationship between a man and a woman: of their desire to be 'one flesh' (Gn. 2: 24); and new in that as time passes, there is the renewal of marriage in Christ, an increasingly modern appreciation of the personal nature and vocation of marriage and the ever fresh choice of the two who want to begin a life together. In the modern religious grasp of human love there is an increasing recognition of what has long been celebrated but, perhaps, poorly recognized: 'Let us be existential; let us see that the love between man and woman is a specific category and type of love, even if we prescind from the sphere of sex, that it is a beautiful and glorious reality that is destined by

---

[1] A slightly different version of this chapter was published by the *Homiletic and Pastoral Review*, 11/08/2014,

http://www.hprweb.com/2014/08/the-holy-family-celibacy-and-marriage-a-reflection-on-the-passage-from-the-jewish-rite-of-marriage-to-the-sacrament-of-marriage/; and, indeed, a slightly different version appeared in Francis Etheredge, *Volume III-Faith is Married Reason*, Newcastle upon Tyne: Cambridge Scholars Publishing, 2016, pp. 182-207.

God's will to play a fundamental role in man's life, and that this love is the classical motive for marriage, that marriage is precisely the fulfilment of this [spousal] love'[2].

There is, however, a negative dimension to the new context for marriage in the situation of "today". On the one hand there is the increasing round-about of relationships where, without planning it, people go from one relationship to another and from one marriage to another; and, in one sense, the "other" person has become the "mirage" of happiness: being happy involves the increasingly unreal expectation that it will come from finding the "right" person. On the other hand, then, losing the meaning of love restores the search for it. Thus there is, as it were, the challenge of love: What is love? What are the dimensions of human happiness? What is it in the turn and turn again to the hope of human happiness that "images" the possibility of a happiness beyond all the disappointments and imperfections of the human person?

Beginning again, then, with an enquiry about the nature of man and woman, marriage and family life is also about beginning from the beginning and yet taking a start from the Christian watershed which re-founds the language of the covenant, the promise of God to do good, in the reality of the Christian sacrament of marriage in which God is present and active: making present the paschal mystery in which Jesus Christ brings new life out of the sufferings and "deaths" of everyday life's humiliations, tragedies and trials. Bioethics is not, then, superimposed upon the reality of human life; rather, just as the human person is a bioethical word[3] so bioethics springs from the deepest account of human life, traversing the problems of meaning and suffering as indeed drawing afresh on the whole reality of an integral account of the human being-in-relationship: to God and to each other.

---

[2] Dietrich von Hildebrand, *The Encyclical Humanae Vitae: A Sign of Contradiction,* Steubenville: Hildebrand Press, 2018, p. 12.

[3] Cf. Etheredge, *The Human Person: A Bioethical Word.*

This first chapter takes us into the transition that brings new life to the mystery of marriage; and, as such, provides a renewable source of energy for the difficulties and sufferings of marriage and family life.

Chapter One: Introduction: "From the Jewish Rite of Marriage to the Sacrament of Marriage". This essay began as a reflection on how, if the Christian Eucharist has its roots in the Jewish Passover, it is possible that there is a Jewish "root" to the Christian sacrament of marriage. It is not a discussion, therefore, about the natural or even the supernatural foundation of marriage, nor does it enter into the "fall" of man and how that impacts on marriage; rather it is about coming to see that the marriage of Mary and Joseph, not always well understood within Christian theology[4], is actually a true marriage which stands at the threshold between the Old and the New Testament[5]: the Old and the New Covenant: the original work of God and how, in Christ, God is not just restoring creation to its original splendour but revealing, more completely, the intimate connection between the mystery of the Blessed Trinity and the mysteries of our salvation. Thus the marriage of Mary and Joseph not only grew in significance as a "moment" in a general movement from Judaism to Christianity but, increasingly, even what seemed like "peripheral" mysteries, such as the perpetual virginity of Mary, Joseph and Christ, transpire to have a central significance, namely that of signifying

---

[4] This confusion about the reality of the marriage of Joseph and Mary, however, may well be a "modern" problem in that the theme of the marriage of Mary and Joseph has been present in liturgical art (conversation with the Rev. Dr. Richard Conrad, Aquinas Institute, 13/08/2014).

[5] Cf. Marc Cardinal Ouellet, *Divine Likeness: Toward a Trinitarian Anthropology of the Family*, translated by Philip Milligan and Linda M. Cicone, Grand Rapids, Michigan/Cambridge, UK: William B. Eerdmans Publishing Company, 2006, p. 190: 'Mary and Joseph nevertheless live a true marriage according to the Old Law, but one which effects the transition to the Kingdom of God inaugurated in the Person and work of their Son.'; and Cf. also pp. 192 and 202.

that marriage is, first of all, a "spiritual act of God", bringing about a union in the context of the salvific communion of God and man.

In the end, then, the coming of Christ is not just incarnational in the sense of God becoming man, as if that was not extraordinary enough; it was, too, incarnational with a view to the transition between an original but fallen gift which God had already begun to renew through the history of salvation and the raising up of all that is good in Jewish culture, culminating in the marriage of Mary and Joseph.

Continuity between the Old and the New Covenant

The Christian sacrament of marriage is "consonant" with that of Judaism and, at the same time, a renewal and development of it[6]. For it seems clear that if the Eucharist "grew" out of the Passover then other sacraments of the Christian Church could have "emerged" out of a similarly significant Old Testament background. Thus the Catholic understanding of the integral nature of Scripture and Tradition (cf. *Dei Verbum,* 7-10; CCC, 75-83[7]) is actually a kind of continuation of the ancient biblical culture of Israel, which was "word", "liturgical rites" and cultural practices. In other words, in the cultural context of the coming of Christ there emerges a sense of an "opportune moment", a favourable time, in which Judaism as a whole and the Jewish rite of marriage in particular was "ripe" for transformation through the coming of Christ. Thus, with the coming of Christ, there is a somewhat seamless passing of the Old Testament rite of marriage, but now bearing a new reality, into the apostolic age of the early Church. As with the

---

[6] This paper has arisen out of giving a number of lectures to a wide range of students on different courses at the Maryvale Institute and, at the same time, given the need for assistance at times of weariness, the need to formalise a kind of dialogue with a number of different sources. I am grateful to those who have contributed, directly or indirectly, to the opportunity to develop these ideas.

[7] *Catechism of the Catholic Church*; hereafter, CCC.

promulgation of the Dogma of the Assumption of Mary, liturgical evidence is a part of the general evidence of the inseparable relationship between Scripture and Tradition[8]; and, therefore, the way that we are introduced to the new nature of marriage *has a kind of discreetness which "fits" in with the logic of the incarnation*[9]. If the sacrament of marriage came to be specified at a particular moment, and the renewal of marriage was itself an indication of a new beginning, "echoing" *the beginning*[10], then it makes sense that the marriage of Mary and Joseph was precisely *the implicitly liturgical moment of that new beginning*.

A New Appreciation of the Transformation of Marriage

It is hoped that these reflections will contribute to the sense that the sacraments, *taking up, as it were, a prior Jewish form,* are revealed in a way that almost indicates a sensitivity to the very "passage" of the transformation of Judaism into Christianity. In other words, marriage stands at the "beginning" of this transformation, taking up as it does the marriage of Joseph and Mary in the context of the coming of Christ, whereas the institution of the Eucharist requires a more explicit beginning. Thus Scripture reflects, as it were, a kind of *implicit* movement, from within, from the Old to the New Covenant: a kind of transition which passes through the natural states of life and suggests a "passage" which is entirely sympathetic to the very states and stages of life so characteristic of our humanity. In other

---

[8] Cf. Pope Pius XII, *Munificentissimus Deus,* 12, 16 and especially 20, where he says: 'since the liturgy of the Church does not engender the Catholic faith, but rather springs from it …'. Thus making clear that the liturgy is not an alternative source of doctrine but, together with Scripture and Tradition as a whole, manifests the fullness of faith and thus can be drawn upon for our understanding of it.

[9] Cf. http://www.rosary-center.org/ll55n6.htm: The Rosary Light & Life - Vol 55, No 6, Nov.-Dec. 2002, Fr. Raftery: the sacraments 'continue' the incarnation.

[10] Cf. St. John Paul II, *Redemptoris Custos,* 7.

words, it may actually be completely relevant to understanding "how" God acts, that the sacrament of marriage is instituted discreetly and recognised as a distinct reality only as the light of Christ's relationship to His Church is understood as the 'marriage of the Lamb' (Rev. 19: 7; cf. also Eph. 5: 31-33)[11].

Background: Scripture and Marriage

In general, the Scripture opens and closes with both the marriage of Adam and Eve, to which Christ refers (cf. Mt. 19: 4-6) and the marriage of the Lamb in the Book of Revelation. Thus the marriage of Adam and Eve constitutes a first and foundational expression of the mystery of marriage; indeed, given the existence of Adam and Eve prior to the Fall through original sin, the union of Adam and Eve fell within the mystery of "original justice" (CCC, 376)[12]. Furthermore, the language of the covenant, applied explicitly after the fall to the promises God made to His people (cf. Malachi, 2: 10), is also applied to marriage (cf. Malachi, 2: 14). Thus, not only is there a development in "interiority" to the explicit covenants themselves, such as between the covenant with Noah and the sign of the rainbow to the covenant with Abraham in the sign of circumcision, but there is also a kind of cross-fertilization between covenant as a relationship between God and His people as a whole and the marriage of a man and his wife. In other words, it becomes increasingly significant that in a Jewish marriage the covenant between God and His people at Sinai is deliberately recalled and constitutes a central

---

[11] I note the help of some comments by Fr. Etienne Nodet, *École Biblique*, who suggested, among other things, the relationship between 'Jesus' celibacy and 'marriage as a vocation' (email, 27 March, 2014).

[12] Although it is not possible to pursue it here, it is interesting to note that death was not a part of the original plan of God and, therefore, marriage was 'forever'. This expression, 'forever', is used in *Familiaris Consortio*, (English translation; CTS edition, 2008): 'that they may remain faithful to each other forever' (20).

Chapter One 55

feature of the marriage ceremony[13]. Thus it is necessary to trace, as it were, some of the interconnected elements that contribute to grasping the growing significance of the relationship between the Jewish rite of marriage and the "institution" of the sacrament of marriage. We are now going to look in a little more detail at a Book which is almost an exposition of marriage.

Thus there are seven "moments" to consider: the Book of Tobit (I); Christ and the Covenant (II); The Marriage of Mary and Joseph (III); Christ and His vocation to Celibacy (IV); Cornelius and his family (V); St. Paul's Letter to the Ephesians (VI); and the early Patristic period (VII).

The Book of Tobit: 2nd – 7th century before Christ[14] (I)

The Book of Tobit is one of the deuterocanonical books, one of the books

---

[13] At http://www.jewishgateway.com/library/rituals/ 19/10/11 Michael Fishbane notes, for example, the relationship between the Jewish marriage ceremony and the remembrance of the marriage of God and Israel at Mount Sinai and the giving of the covenant. Furthermore, a ring is used in the traditional ceremony: "You are betrothed to me, with this ring, in accordance with the laws of Moses and Israel". Nevertheless, as a conversation with Dr. Andrew Beards (at the Maryvale Institute at the time) brings to light (18/10/11), there must be a certain caution about establishing what was the practice and understanding of Jewish marriage at the time of Christ, Nevertheless, it is necessary to give this reflection on the Jewish rite of marriage and the Christian sacrament of marriage some scholarly foundation. Even so, however, it is clearly of some significance that Christ was present at an actual marriage, quite apart from the "symbolism" of such an event in terms of the significance of St. John's theology of Christ's mission (cf. Jn. 2: 1-11). In other words, there is a meaning to investigate in both the literal and spiritual sense (cf. CCC, 116-117) of this text.

[14] Tobit, dated earlier by modern scholars but there is no reason given here, p. 920 of the *Catholic Bible Dictionary*, General Editor: Scott Hahn, Doubleday: New York, First Edition, 2009; indeed, Shalmaneser Vth was reputedly the one mentioned in the Book of Tobit, Shalmaneser, p. 833.

disputed as to whether or not it was Scripture: 'Rabbinic Judaism and founders of Protestantism rejected the deuterocanonical books; some Protestant Bibles print them in a separate section called "Apocrypha"'[15]. One wonders what determined this rejection of the Book of Tobit; and, as such, one answer was that its authorship was so late, reputedly '100 AD', although new evidence dates it significantly earlier[16]. Another reason for its rejection was that it expressly contravened a rabbinic practice: 'Other scholars have

---

[15] *Catholic Bible Dictionary*, General Editor: Scott Hahn, Deuterocanonical, pp. 213-214; and, as such, deuterocanonical comes 'from the Greek for "second canon"' and was first used in 1569. 'Books regarded as canonical with little or no debate were called "protocanonical" (from the Greek for "first canon")' p. 213. The Apocryphal Books: Greek for "hidden things" and applied by the Catholic Church to books 'that are often similar to the inspired works in the Bible' but were not judged to be part of the canon (Catholic Bible Dictionary, General Editor: Scott Hahn, Apocryphal Books, p. 54). The Catholic Church recognised that Tobit, and the other books, belonged in the Canon of Scripture as enumerated in 1546 (Catholic Bible Dictionary, General Editor: Scott Hahn, Deuterocanonical, p. 214). Even if, then, there is some dispute as to its age and it is rejected by some, the Book of Tobit is nevertheless a part of a considerable body of references throughout the whole of Scripture and Jewish tradition to marriage (consider, for example, the article in Encyclopaedia Judaica, Jerusalem: Keter Publishing House Ltd., 1971: Volume II: LEK-MIL, pp. 1026-1054; and, in addition, Catholic Bible Dictionary, General Editor: Scott Hahn, Marriage, pp. 577-582).

[16] http://en.wikipedia.org/wiki/Book_of_Tobit: 'Prior to the 1952 discovery of Aramaic and Hebrew fragments of Tobit among the Dead Sea Scrolls in Cave IV at Qumran, it was believed that Tobit was not included in the Jewish canon because of its late authorship, which was estimated to be circa 100 AD. However, the Qumran fragments, which date from 100 BC to 25 AD and are in agreement with the Greek text existing in three different recensions, evidence a much earlier origin than previously thought.'

postulated that Tobit was excluded from the Jewish Scriptures for a halakhic[17] reason, because the marriage document discussed in 7:16 was written by Raguel, the bride's father, rather than by the groom, as required under Jewish rabbinical law'[18]. While it is not possible to discuss this further, it is clearly significant that the Book of Tobit is a controversial text and, as such, stands so clearly in the development of our understanding of marriage. It has also been noted that in this book there is mention for the 'first time ... in the Bible [of] a formal marriage contract involving a written document'[19]; however, it is mentioned in such a way as to make one think that this was a usual practice and was, 'in time ... [to] be called by Jews the *Ketubah*'[20].

The Book of Tobit

In general the *Book of Tobit* depicts marriage at two stages: a marriage and its difficulties towards the end of life (Tobit and his wife Anna); and, at the same time, considers the difficulties that beset the possibility of marriage (Tobias, the son of Tobit and Anna and Sarah the daughter of Raguel and Edna). There is an angel, Raphael, who both leads Tobias to Raguel and to marriage with his daughter Sarah; and, at the same time as the angel instructs Tobias on how to defend Sarah from a demon (Tobit, 6: 7), Raphael also

---

[17] http://en.wikipedia.org/wiki/Halakhic: '*Halakha* has been developed and pored over throughout the generations since before 500 BCE, in a constantly expanding collection of religious literature consolidated in the Talmud. First and foremost it forms a body of intricate judicial opinions, legislation, customs, and recommendations, many of them passed down over the centuries, and an assortment of ingrained behaviors, relayed to successive generations from the moment a child begins to speak. It is also the subject of intense study in yeshivas; see Torah study.'

[18] http://en.wikipedia.org/wiki/Book_of_Tobit

[19] *The Navarre Bible: Chronicles – Maccabees,* General Editor, Jose Maria Casciaro, Dublin: Four Courts Press, 2003, p. 324.

[20] *The Navarre Bible: Chronicles – Maccabees,* p. 324.

instructs Tobias on how to cure his father's blindness (Tobit, 6: 8).

Although there is a sense in which Tobias is marrying one to whom he is, as it were, intended by kinship (Tobit, 4: 13; 6: 9-17), Sarah's marriage to Tobias is nevertheless in answer to her own prayer for deliverance from this demon and her own unhappiness (cf. Tobit, 3: 10-17). Furthermore, lest we think in terms of Sarah's readiness to marry someone she does not know except in terms of how he has presented himself to her father and the family Tobias comes from, we can already see that Tobias fell in love with Sarah even before he met her; for, following the angel's description of her and of how to help her and to conduct himself, it says of Tobias that 'he fell in love with her and yearned for her deeply' (Tobit, 6: 17). The key question, then, is not so much the modern and contentious question of "individual freedom"[21] but, rather, Tobias and Sarah's participation in a dialogue with God and, therefore, with a perception of their own happiness in the context of the "providential" action of God at work in their own lives.

On meeting Raguel, Sarah's father, Tobias declares his desire to marry Sarah to Raphael and Raguel overhears and, at Tobias' insistence, makes a "gift" of his daughter Sarah to Tobias: taking Sarah by the hand, Raguel gave her to Tobias saying 'take her to be your wife in accordance with the law and the decree written in the book of Moses' (Tobit, 7: 13; and 7: 1-15); and then afterwards, Raguel asks his wife Edna 'to bring writing material; and he wrote out a copy of the marriage contract, to the effect that he gave her to him as wife according to the law of Moses' (Tobit, 7: 14).

Finally, Tobias followed the instructions of Raphael and drove the demon away from Sarah. Tobias and Sarah then prayed together and slept; and, beginning the following day, there was a two week celebration of the

---

[21] In "Marriage and Divorce" by Michael L. Satlow, there is clearly an acknowledgement of the modern question of feminism and the rights of women, pp. 4 and 6; however, the question arises about the action of providence as understood and mediated by events for the benefit of both the man and the woman.

marriage, twice as long as usual[22], and a gift of half of Raguel's family wealth to Tobias (cf. Tobit, 8). In passing, we can note, too, that the prayer that Tobias makes refers to *the beginning* in these words: 'You made Adam and gave him Eve his wife as a helper and support. From them the race of mankind has sprung' (Tobit, 8: 6); and if we consider all the elements of the religious life which are practised (almsgiving; prayer; trust in providence etc.) and which, therefore, support marriage, we can see what a comprehensive catechesis *on the integrity of the life of faith* is embodied in the Book of Tobit.

The Covenant of God and Israel: Christ and the Covenant (II)

Given the key relationship, for Israel, of the covenant of Sinai as a kind of marriage between Israel and God, it brings a new significance to the marriage feast of Cana (Jn. 2: 1-11[23]) to find that Christ is present at what is, according to traditional Judaism, a "moment" of remembrance of God 'wedding ... Israel'[24]; indeed, Christ Himself, in referring to the gift of the Eucharist, takes up precisely this language of the covenant, particularly at the Institution of the Eucharist when He refers to the cup, 'saying, "This cup which is poured out for you is the new covenant in my blood ..."' (Lk. 22: 20). Thus the Church says of Christ's presence at Cana that it is both the 'confirmation of the goodness of marriage and the proclamation that henceforth marriage will be

---

[22] *Navarre Bible: Chronicles – Maccabees*, p. 327.

[23] *Catholic Bible Dictionary*, General Editor: Scott Hahn, John, Gospel of, pp. 459-460: Although there is dispute about authorship, there is both historical evidence and evidence from tradition that St. John was the beloved disciple who wrote this Gospel, even as early or earlier than AD 70; for example, owing to a reference to the 'Sheep Gate', 'as though the city [of Jerusalem] was still intact at the time of writing' and thus suggesting the Gospel was written before Jerusalem was destroyed in AD 70.

[24] http://www.jewishgateway.com/library/rituals/ (24/10/11).

an efficacious sign of Christ's presence' (CCC, 1613); indeed, 'This grace of Christian marriage is a fruit of Christ's cross, the source of all Christian life' (CCC, 1615). Thus the Catechism indicates the *perspective of the cross*, as the perspective in which the sacrament of marriage will come to exist: the cross through which the water of suffering is changed into the wine of rejoicing and thanksgiving (cf. Jn. 2: 3-10).

The Gospel of St. Matthew[25], a Gospel that was thought to be 'written for Jewish Christians' as it 'stresses that Christ is the fulfilment of the Old Testament Scriptures' and 'Jewish customs are mentioned without explanation', is traditionally ascribed to the apostle Matthew[26]. In this particularly Jewish Gospel, a Gospel which comes first in the New Testament canon and, as such, is almost an indication of the Jewish roots of Christian life, we find a number of references to marriage. In the details of the parables, such as the virgins carrying lamps (cf. Mt. 25: 1-12[27]), in reference to wedding garments (cf. Mt. 22: 12[28]) and a feast (cf. Mt. 22: 2-3; 25: 10[29]); and, therefore, there is the general impression that the parables are drawn from familiar and contemporary experience.

---

[25] *Catholic Bible Dictionary*, General Editor: Scott Hahn, Matthew, Gospel of, p. 591: again reputedly written before the destruction of Jerusalem, such that 'Suggested dates for the book have thus ranged from A.D. 50 to 100'.

[26] *Catholic Bible Dictionary*, General Editor: Scott Hahn, Matthew, Gospel of, p. 590, and 591.

[27] Cf. *Encyclopaedia Judaica*, Marriage, p. 1042.

[28] Cf. *Encyclopaedia Judaica*, Marriage, p. 1033, the bride and groom and by implication all others.

[29] *Encyclopaedia Judaica*, Marriage, p. 1032.

The Cultural context of the Marriage of Mary and Joseph (III)

When it comes to the marriage of Mary and Joseph, there is no mention of the custom of arranged marriages nor of the arrangements that surround the marriage of Mary and Joseph; however, as with the marriage of Tobias and Sarah, it seems possible to consider these things in the context of prayer and the answering action of God in which, in general, the devout understood their lives to be lived in the context of divine providence. There is an indirect mention, as we shall see, of a distinction between a legally binding act of betrothal, called *qiddushin* (pronounced *kiddushin*) and, in due course, the married couple coming to live together[30].

A question that arises, then, is this: if this distinction between a binding betrothal and cohabitation was a relatively unique feature of Jewish marriage in the 'ancient Near East'[31] – then what prompted its development? Could the

---

[30] In "Marriage and Divorce" by Michael L. Satlow, p. 4; however, as with Scripture generally, there is a certain amount of controversy about these practices e.g. what is the relationship 'of this law to lived experience'? (p. 4). Nevertheless, the practice was obviously settled enough, from ancient times, for Goldberg to say 'While this act [*kiddushin*, consecration, or *erusin*, betrothal], establishes the legal bond of marriage, until the late Middle Ages bride and bridegroom generally went on living in their respective homes for another year. Only then would the marriage be consummated. In ancient times *chuppah* was the name given to the hut or chamber in which the consummation took place' (Goldberg and John D. Rayner, *The Jewish People: Their History and Their Religion*, London: Penguin Books, 1989, p. p. 375).

[31] In "Marriage and Divorce" by Michael L. Satlow, p. 8, although later, in this same article Satlow says: "There is now wide (but not unanimous) scholarly agreement that there was little that was substantively and significantly different about Jewish marital and divorce practices in antiquity.' This does not make sense if we are to understand a connection, implicit and explicit, to the Covenant between God and His people. Furthermore, while there may be some evidence of similarity between Jewish and non-Jewish law (cf. Satlow, pp. 5-6 of "Marriage and Divorce"),

answer be that it was a "ritual" development of the growing understanding, within Judaism itself, of the significance for marriage of the Covenant at Sinai? For example, it is the prophet Malachi who uses the language of the covenant for both marriage and the relationship of God to Israel; and, if Malachi was about 'the fifth century B.C.,'[32] then it puts this development almost between the Book of Tobit and the New Testament.

What complicates the question of the history and origin of the distinction between *betrothal* and *coming to live together*, however, are the two, if not more references to this practice in the books of Exodus and Deuteronomy. Taking them canonically, in the sense of where they occur in the canonical text, the first comes in the context of the 'Violation of a virgin' and indicates, indirectly, the betrothal of a virgin: "If a man seduces a virgin who is not betrothed ...' (Ex. 22: 8[33]). Furthermore, in the case of the latter text from the

---

the very existence of the New Testament evidence (cf. Mt 1: 18-25) and its subsequent rabbinic elaboration (cf. *Encyclopaedia Judaica*, Marriage, p. 1032; and cf. also http://www.jewishencyclopedia.com/articles/14216-talmudic-law), speaks of both a real cultic practice and its enduring through time. In other words, there arises the question: why did this distinction between betrothal and cohabitation come to exist at all?

[32] *Catholic Bible Dictionary*, General Editor: Scott Hahn, Malachi, Book Of, p. 567.

[33] The text and reference to any commentary, for the verses referred to here, are taken from *The Navarre Bible: Pentateuch*, J. M. Casciaro, Director of the Editorial Committee, Dublin: Four Courts Press and Princeton, NJ: Scepter Publishers, 1999. The commentary, at this point, does not particularly note the reference to the practice of betrothal before coming together; however, it does mention the following: 'a marriage ceremony was a commitment made between families and that the family gaining this new couple in some way compensated the family which was deprived of it' (part of the commentary on Ex. 22: 16-17, p. 339). The salient point here, for the sake of this discussion, is the indication of 'a marriage ceremony'; and it is relevant because it indicates a distinction, perhaps too obvious to mention, between 'a

book of Exodus, there is the immensely significant context of the delivery of the people from their slavery in Egypt and the "theophonic" giving of the Ten Commandments on Mount Sinai. What is more, the legislation which includes the reference to 'a virgin who is not betrothed' (Ex. 22: 8) is a part of the legislation that *constitutes the Covenant of Sinai* (Ex. 20: 22 – 23: 33), which is made very clear by the heading from which it begins, namely that it is 'The Book of the Covenant'[34]. Thus, by implication, there is a very clear relationship between the Covenant at Sinai and the distinction between *betrothal* and *coming to live together*. The second text, referring to soldiers who can go back from the battle front when confronting an army 'larger than your own' (Dt. 20: 1), again refers to betrothal, but this time more directly to the difference between betrothal and coming together: 'And what man is there that has betrothed a wife and has not taken her? Let him go back to his

---

ceremony' and a subsequent coming together. However, the many other and perhaps more significant elements of this passage cannot be addressed here.

[34] Heading 'B. The Book of the Covenant', p. 103, which falls between Ex 20: 21 and Ex. 20: 22, of *The Jerusalem Bible,* general editor, Alexander Jones, Philippine Bible Society: Darton, Longman and Todd, 1966. Cf. p. 103, footnote 20h of *The Jerusalem Bible* and p. 331 of the *Pentateuch,* but the discussion cannot be pursed here concerning the origin of these laws, suffice it to say that the overriding implication is that, as both commentators acknowledge, these laws fall within what is 'sanctioned by God himself and as part of the obligations of the Covenant' (p. 331 of *Pentateuch*; cf. also, for example, the witness of Scripture itself: 'He declares his word to Jacob, his statutes and ordinances to Israel' Ps. [147]: 19). Moreover there is the additional significance of the following section entailing the celebration of a meal in the presence of God, known as 'The Covenant is Ratified' (falling between Ex. 23: 33 and Ex. 24: 1). One can only mention in passing, too, the whole tradition of the "marriage" of God and His people and, indeed, the "opening", public presence and first miracle of Christ at the marriage feast of Cana.

house ...' (Dt. 20: 7³⁵)³⁶. Thus there is a kind of indication that the reality of the distinction between betrothal and coming together may, possibly, be established ceremonially and, therefore, liturgically. Nevertheless, we cannot strictly rule out the use of a written document for marriage in so far as there was a written document for divorce (although divorce itself was repudiated by Christ)³⁷.

With respect to the Book of Tobit, however, while there is no explicit use of the word 'covenant', it is interesting to note that Sarah's father, Raguel, when drawing up the 'marriage contract' (Tobit, 7: 14), 'wrote out a copy of the marriage contract, to the effect that he gave her [Sarah] to him [Tobias] as wife according to the law of Moses' (Tobit, 7: 14). Thus, while there is no explicit use of the word covenant, it seems there is clearly an understanding that marriage is specified by the law of Moses and, as such, by the context, then, of the covenant between God and Israel. Incidentally, in a translation of the Book of Tobit from the Greek, the translation of this line says: 'And called Edna his wife, and took paper, and did write an instrument of covenants, and sealed it'³⁸. Thus there is a suggestion of the language of the

---

[35] There is a twofold reference here. The first is to a cross reference to Dt. 24: 5, (on p. 744 of *Pentateuch*) which refers to a newly married man not going out with the army; and the second reference, this time to a part of the commentary on Dt. 20: 1-20, makes explicit the 'two phases (betrothal and bringing the betrothed woman to her husband's house)' (p. 743 of *Pentateuch*) of marriage.

[36] There are two other mentions of a betrothed virgin. In the first one, Dt. 22: 23-24, both are killed for their breach of marriage; and in the second one, there is an injustice to the woman and, this time, the response of death to the man who perpetrates it (Dt. 22: 25). In the case of the commentary on Dt 22: 25, the commentary adds 'A "betrothed" woman ... already had the same obligations to faithfulness as a wife' (pp. 749-750 of *Pentateuch*). There is a final reference to a woman not being 'betrothed' at Dt. 22: 28.

[37] Cf. the commentary to Dt. 24: 1-4, pp. 754-756 of *Pentateuch*.

[38] ellopos.net/elpenor/greektexts/septuagint/chapter.asp?book=18&p age=7

covenant being present in the Greek text, although not as a form of the usual noun, διαθηκη³⁹, but in the range of meaning attributed to the word συγγραφην⁴⁰. In addition, there is the whole emphasis, as it were, in the book as a whole, on the two tables of the law: the love of God and the love of neighbour, with particular reference to the 'fourth commandment [that] opens the second table of the Decalogue'[41].

In brief, although it is not so obviously the case with the marriage of Tobias and Sarah, was there "already" a development of the parallel between the giving of the Law at Sinai and the people dwelling with God in the desert, and the distinction in the rite of marriage between the formal betrothal and the later coming together: a kind of ritual expression of the relationship between the marriage of a man and a woman and the marriage of God and His people[42]? Furthermore, does the early distinction between betrothal and coming together also indicate, in the very fact of this distinction, the analogous distinction in the Catholic understanding of sacrament: an action of God and the consequence of this expressed by the couple coming together? Let us note, in this context, the words of the prophet Malachi, cited by the commentary on Deuteronomy (Dt. 24: 1-40), where the prophet says: 'Because the Lord was witness to the covenant between you and the wife of your youth, to whom you have been faithless, though she is your wife by

---

[39] Other translations, however, do not use the word covenant e.g. http://ccat.sas.upenn.edu/nets/edition/19-tobit-nets.pdf

[40] The Septuagint Greek of the Book of Tobit (BibleWorks 6).

[41] *Navarre Bible: Chronicles – Maccabees*, p. 330, where the commentator is quoting from the *Catechism of the Catholic Church* (CCC), 2197.

[42] Although this speculation is not based on a particular source, as such, it is nevertheless the case that many Jewish marital practices are a kind of liturgical expression of biblically rooted "history"; for example, the elements of the 'chuppah' or the practice of the bride and groom being under a canopy 'outside, under the stars, as a sign of the blessing given by G-d to the patriarch Abraham' (http://ohr.edu/1087 : "The Jewish Wedding Ceremony" by Rabbi Mordechai Becher.

covenant .... For I hate divorce, says the Lord the God of Israel [...]'[43]. Furthermore, then, could there be a progress, from the time of the Book of Exodus, from a liturgical to a liturgical and a written contract of marriage; and, as such, a growing recognition of the significance of the distinction between being betrothed and coming together?[44] In other words, could the growing recognition of the distinction between betrothal and coming together signify a deeper understanding of the "divine-human act" of betrothal: a significance that both fully involves the humanity of the man and woman and, at the same time, entails a deeper awareness of how they are "embedded" in the history of salvation?[45] Finally, there is another consideration which, while not particularly explicit in these particular texts or commentary, is that emphasising the distinction between betrothal and coming together also allows for a period of adjustment and preparation for the bride, if not for the groom, too.

Pope St. John Paul II and Redemptoris Custos

Thus, in the letter of St. John Paul II on St. Joseph, *Redemptoris Custos*, we see that the Pope reflects on Scripture drawing on a tradition of the distinction between marriage and cohabitation, both contemporary to and indicated in the marriage of Joseph and Mary in St. Matthew's Gospel (cf.

---

[43] *Pentateuch*, p. 756, the whole extract coming from 'Malachi 2: 13-16' which, as it says on p. 755, 'is the clearest and most forceful Old Testament passage to condemn divorce, on the grounds of the religious character of marriage: it is something which is similar to God's Covenant with Israel'.

[44] It is not possible, however, to discuss all the variations in the law and the issues raised by different judgments; for example, cf. Dt. 22: 28-29 and the comment on p. 749 of *Pentateuch* e.g. the different sanctions applied to a man and a woman in the situation described at Dt. 22: 13-30.

[45] Cf. Dt. 38: 1-27 and pp. 184-187.

Mt. 1: 18-21; *Redemptoris Custos*, 18[46]). Furthermore, St. John Paul II draws on Blessed Pope Paul VI who said: "In this great undertaking which is the renewal of all things in Christ, marriage - it too purified and renewed - becomes a new reality, *a sacrament of the New Covenant*[47]. We see that at the beginning of the New Testament, as at the beginning of the Old, there is a married couple. But whereas Adam and Eve were the source of evil which was unleashed on the world, Joseph and Mary are the summit from which holiness spreads all over the earth. The Savior began the work of salvation by this virginal and holy union, wherein is manifested his all-powerful will to purify and sanctify the family – that sanctuary of love and cradle of life" (*Redemptoris Custos*, 7).

Although one possible beginning to the sacrament of marriage is Christ's presence at the marriage feast of Cana[48], in the very complex mystery of the Holy Family, we have grounds for understanding that the marriage of Mary and Joseph was in fact the first marriage: a sacrament of the New Covenant. In other words, through the mysteries present in this unique event, Mary, who was conceived without sin (Lk. 1: 28), has already received, as it were, the "baptism" of being the Immaculate Conception; secondly, on believing the message of the angel that she was to conceive 'Jesus' (Lk. 1: 31) and become the mother of the 'Son of the Most High' (Lk. 1: 32), she has professed her faith in Christ.

When Joseph, her husband, believed the angel that the child Mary has conceived is 'of the Holy Spirit' ... and you shall call his name Jesus, for he will save his people from their sins' (Mt. 1: 20-21), then it is clear that Joseph has undergone a "conversion". Joseph has been turned from being willing to send Mary away quietly, having married her and found that she is pregnant

---

[46] *Encyclopaedia Judaica*, Marriage, p. 1032 etc.

[47] Emphasis added.

[48] *Catholic Bible Dictionary*, General Editor: Scott Hahn, Marriage, p. 580: 'Traditional exegesis often states that Christ sanctified marriage by his presence at the Cana feast and by his contribution to the festivities (i.e., the good wine).'

but not by him, to being able to take her to his home. Thus Joseph, too, has implicitly professed his faith in Jesus; and, although there is no mention in Joseph's case of any kind of "baptism", yet it is possible that in Joseph's conversion to Christ there is an action of God equivalent to baptism: 'Since it is inconceivable that such a sublime task would not be matched by the necessary qualities to adequately fulfil it, we must recognize that Joseph showed Jesus "by a special gift from heaven, all the natural love, all the affectionate solicitude that a father's heart can know"(26[49])' (*Redemptoris Custos*, 8). In other words, both Mary and Joseph seem to be replete with all the ingredients of what constitutes Christian marriage; and, what is more, in so doing their lives seem to reflect an implicit participation in all the graces that are a part of it, explicitly so in the case of Mary's Immaculate Conception being a "baptism".

Furthermore, then, in that their child is Jesus Christ, true man and true God and the new high priest (cf. Heb. 4: 14 - 5: 1 10; and 6:11 - 10: 18[50]), it is as if we can say that the Holy Family is replete with all the sacraments of the Church, either explicitly or implicitly. On the one hand, perhaps we do not disclose anything by the claim that the whole sacramental order was "inchoately" present in the Holy Family, precisely because it was not explicit. On the other hand, if existence comes before thought, there is a sense in which the actual reality of the Holy Family may have a certain priority and "governing" influence over the development of doctrine: 'it is in the Holy Family, the original "Church in miniature (*Ecclesia domestica*)," (19) that every Christian family must be reflected' (*Redemptoris Custos*, 7).

Thus, it is not so much that there needs to be a discreet institution of the

---

[49] Footnote 26 of *Redemptoris Custos* shows that St. John Paul II is quoting from Pius XII: '26. Pius XII, Radio Message to Catholic School Students in the United States of America (February 19, 1958): AAS 50 (1958), p.174.'

[50] Cf. A reading from the "Evangelical commentary" of Eusebius of Caesarea, bishop, p. 174 of Readings for Lent and Paschal Time, Volume II, *Pro Manuscripto*: For private circulation only.

actual sacraments, as the actual sacraments are an unfolding of the implicit *logic of the incarnation as it unfolded in the life of Mary, Joseph and Jesus*. Although there does not need to be a discreet institution of the sacraments one can see, nevertheless, that there are "moments" when what is "implicit" in the mystery of salvation is made "explicit"; and, to that extent, there is both an unfolding of *the logic of the graces given to Mary and Joseph through the coming of Christ* and there are particular moments when the sacramental mystery, entailed in the very nature of the Holy Family's existence in Christ, are made manifest. A clear example of the latter is, of course, the Institution of the Eucharist (cf. Mt. 26: 26-29). On the one hand, there is the very mystery of Christ being God made man; and, on the other hand, Christ takes up "from within" the tradition of the Passover and the offering of the paschal lamb (cf. Jn. 1: 29, 36). Furthermore, Christ *translates,* as it were, the paschal gift of His self-offering into the gift of the 'bread of life' (Jn. 6: 35).

The interrelationship of Christ's 'chosen celibacy' and the vocation to marriage[51] (IVi)

A further question that arises, then, is precisely the emergence of the understanding of Christ's love of the Church as a development which assisted in the articulation of the sacrament of marriage. In other words, in general, it appears that just as *the ecclesiology of the Church as a whole was the natural context for the Second Vatican Council referring to the domestic Church* (*Lumen Gentium,* 11), so our understanding of the sacrament of marriage is "founded" on a clear perception of Christ's love of the Church. On the one hand there is the supposition of the complete and *integral nature of the gift of grace to the Holy Family of Nazareth: Jesus, Mary and Joseph*; and on the other hand there is the canonical structure of the New Testament, beginning with the Gospel of St. Matthew and ending with the Book of

---

[51] Fr. Etienne Nodet, email, 27/3/14.

Revelation. Thus it is almost as if the New Testament as a whole is a passage from a consciousness of the Jewish inheritance, as epitomised in the Gospel of St. Matthew and never, in a sense, superseded, to a complementary grasp of the spiritual realities unfolded both in the New Testament and also in the Scripture as a whole.

The Gospel of St. John, opening with the meditation on the 'Word', (the Logos, λογοσ) which 'became flesh' (Jn. 1: 14), comes as the last of the Gospels, fittingly bridging the transition to the uniquely spiritual understanding, culminating in the Book of Revelation, of Christ as the bridegroom of the Church. On the one hand, not only are there many references in the Gospels to marriage, but John the Baptist applies this imagery to Christ Himself: 'the friend of the bridegroom, who stands and hears him, rejoices greatly at the bridegroom's voice' (Jn. 3: 29)[52]. On the other hand there is a sense in which these references are "understood" to apply both to a great wedding feast (Mt. 22: 1-14), implying the many who are coming, and then also Christ applies the image of the bridegroom to Himself: "The kingdom of heaven may be compared to a king who gave a marriage feast for his son ..." (Mt. 22: 2). As a whole, then, not only does the New Testament go from the event of marriage, in particular the marriage of Mary and Joseph, to the spiritual significance of that marriage-event, but the Scripture as a whole goes from the marriage of Adam and Eve to the marriage of the Lamb[53].

---

[52] It is interesting to note that Michael L. Satlow, in *Jewish Marriage in Antiquity*, Princeton University Press, 2001, Chapter 1 published as a pdf on the web, says of Christ: 'Where Jesus does evoke the image of marriage, the "marriage" is between himself (the groom) and those who anticipate the kingdom of God' (citing a reference, 139), p. 24.

[53] 'St. Thomas Aquinas says there are two things which can order time and sense in Scripture. The first is that "human minds, existing in bodies, know first the natures of material things, and by knowing the natures of what they see, derive some knowledge of what they cannot see" (*STh* I. 84. 7); and the second is that "nothing

There is an emergence of Christ's vocation to celibacy[54] in the context, then, of the Holy Family of Nazareth, in which all things seem to be inchoately present *in terms of the "reality-sign" of the interrelationship between marriage, the vocation to virginity and the unfolding of family life.* Thus the vocational celibacy of Mary and Joseph is in the context of their marriage which, in its own way, seems to give grounds for *the vocational*

---

necessary to faith is contained under the spiritual sense which is not elsewhere put forward by the Scripture in its literal sense" (*STh* I. 1. 10)' (excerpt from an article by the author: http://www.hprweb.com/2012/01/scripture-is-a-unique-word/).

[54] A consideration of the relationship between Christ's vocation to celibacy and the Essenes (Fr. Etienne Nodet, email, 27/3/14), although another line of enquiry, will have to suffice with the comment that if celibacy was beginning to show itself in a new and striking way, that this was a part of the general making ready for the coming of Christ which constitutes a part of this being the providential time. According to Satlow, for example, in *Jewish Marriage in Antiquity*, Chapter 1, p. 24: the Dead Sea community moved toward an 'eschatological stance' wherein 'procreative couples continue to be necessary in the here and now, but families will lose relevance in this next stage of history.' The evidence reviewed by Satlow entails the possibility of both married and celibate members of this community and a certain emphasis on the priesthood (cf. p. 24). In a more particular sense, then, there remain questions to be investigated; for example, although the relationship between John the Baptist and a tradition of Old Testament priesthood is well established (cf. Pope Benedict, *Jesus of Nazareth: The Infancy Narratives*, translated by P. J. Whitmore, London: Bloomsbury, 2012, pp. 18-23), there is the question of how 'the priesthood of the Old Covenant [epitomised, in a sense, in John the Baptist] moves toward Jesus' (p. 18)? In one sense, Benedict begins to answer this question by speaking of Joseph as 'son of David' and, as such, 'he is to bear witness to God's faithfulness' (p. 42). Nevertheless, given that the fullness of priesthood is, in a sense, what is unique to Christ (cf. for example, *The Navarre Bible: Hebrews: Texts and Commentaries*, translated by Michael Adams, Dublin: Four Courts Press, 1991, pp. 90-102), there are clearly many questions to be answered about the development of His self-understanding. In particular, how is the self-understanding of Christ's priesthood "assisted" through His family life?

celibacy of Christ being understood in the light of marriage: the spiritual marriage between God and His people. For just as the spiritual marriage between God and His people is *embodied, as it were, in the vocational celibacy of both Mary and Joseph* so, conversely, Christ may have understood His vocational celibacy in the light of both the Covenant of Sinai and the covenant of marriage between Mary and Joseph. Moreover, in the light of what Pope John Paul II says in *Familiaris Consortio,* the interrelationship of love and fecundity in the Holy Family of Nazareth seems to give rise to a particularly extraordinary image of the interrelationship between the Holy Family and the communion of the Blessed Trinity: 'God is love and in Himself He lives a mystery of personal loving communion. Creating the human race in His own image and continually keeping it in being, God inscribed in the humanity of man and woman the vocation, and thus the capacity and responsibility, of love and communion. Love is therefore the fundamental and innate vocation of every human being' (11[55]).

While, then, there are other factors in the dialogue between the vocations to marriage and celibacy and the articulation of their respective differences and similarities, *the unique nature of the Jewish marriage and family life of Mary, Joseph and Jesus* is clearly a contributory factor *in the inner logic of Christ understanding His relationship to the people of God in the light of God's covenantal love of His people.*

---

[55] This quotation has excluded the footnotes in the original.

Chapter One

The word of God, Christ and the vocation to celibacy (IVii)

Drawing on a background of ideas and practices[56] that do not so much reflect the exact times of the Holy Family as help us, nevertheless, to see that there are trends in thinking which were probably older and, as such, could have a bearing on how to understand a key relationship in the life of Christ, namely, that between the "lived Scriptures" of His daily life and worship and the mystery of His own identity.

On the one hand there is the very mystery of the Word: the 'Word became flesh' (Jn. 1: 14); and, therefore, the mystery of the Second Person of the Blessed Trinity being manifest in the person and activity[57] of Jesus Christ. On the other hand Jesus was subject to His supposed father, Joseph: 'feeding, clothing and educating him in the Law and in a trade, in keeping with the duties of a father' (*Redemptoris Custos*, 16)[58]. Moreover, 'marriage ...

---

[56] Satlow, *Jewish Marriage in Antiquity*, Chapter 1: In the course of reading Matthew Satlow's opening chapter, which entails setting a part of the Babylonian Talmud's discussion on marriage in its cultural context, there emerge a number of considerations that help us to understand a new relationship between the Holy Family and Christ's vocation to celibacy. In the first place, however, this is clearly not the purpose of Satlow's own study; rather, he has in mind, as I say, the contextualization of "two" strands of thought in the Babylonian Talmud: a Palestinian emphasis on marriage as founding a household (an οικοσ), the public good of this institution and that this is to act in accord with human nature (pp. 12-21); and a Babylonian emphasis on marriage as having a 'covenantal significance' (p. 20), which entails understanding marriage as positively ordered to causing 'God to dwell amidst Israel, and can even bring the Messiah' (p. 28), as a remedy to sinful inclinations and, therefore, to assist in the disposition necessary for men to study the Torah (p. 30).

[57] Cf. The scholastic adage: activity manifests being.

[58] Note also that the Catechism has included a section on the early life of Christ under the heading: 'II. The Mysteries of Jesus' Infancy and Hidden Life' (CCC, 522-534), an innovation note by Cardinal Avery Dulles: 'Although the creed skips from

establishes the family as the basic social unit and the home as 'the little sanctuary' (Ezekiel, 11: 16) in which the father is like a priest, the mother like a priestess, and the dining-room table like an altar (Berachot 55a), where children can enjoy their childhood and grow to maturity under the loving protection and guidance of their parents, and where the Jewish religion can be practised, experienced and transmitted from generation to generation'[59]. Thus Joseph fulfils the injunction of the *Shema*[60] as regards his child: 'and you shall teach them diligently to your children' (Dt. 6: 7); and what Joseph shall teach his child Jesus is that 'The Lord our God is one LORD; and you shall love the LORD your God with all your heart, and with all your soul, and with all your might' (Dt. 6: 4-5). Although it is not clearly established that Jesus had observed the *bar mitzvah*, the rite that establishes a Jewish boy of thirteen as responsible for fulfilling the law himself[61], in the answer that Jesus gave to His parents when He was twelve, on being found by them in the Temple[62], Jesus shows that He has identified His mission with that of His Father's house and, in a sense, expressed a wholehearted love of God and

---

the birth of Jesus to his passion and death, the Catechism inserts at this point a section on the mysteries of the life of Jesus' (p. 49, "The Challenge of the Catechism", *First Things,* no 49, Jan, 1995, pp. 46-53).

[59] Goldberg, *The Jewish People: Their History and Their Religion*, p. 371.

[60] From the Hebrew word for 'Listen' (Dt. 6: 4); and, in general, I am indebted to a certain sensitivity to this word being applied to Christ through the liturgies and catecheses of the *Neocatechumenal Way* and being a particularly significant prayer in Judaism (cf.

http://en.wikipedia.org/wiki/Shema_Yisrael)

[61] http://www.jewfaq.org/barmitz.htm: can be one of the 'the minimum number of people needed to perform certain parts of religious services, to form binding contracts, to testify before religious courts and to marry.'

[62] biblegateway.com/resources/commentaries/IVP-NT/Luke/Twelve-Year-Old-Jesus-Goes where there is the point made that: 'Jesus' parents--and Luke's readers--need to appreciate that Jesus understood his mission' (with reference to Lk. 2: 41-51).

dedication to His will: "Did you not know that I must be in my Father's house?" (Lk. 2: 49). In other words Joseph has accomplished his mission as a father in assisting Jesus to understand the history of salvation as the *Shema* commands; and, in a certain way, the boy Jesus, through his public expression of who He is to His parents, has fulfilled the *bar mitzvah* and is now able 'to perform certain parts of religious services, to form binding contracts, to testify before religious courts and to marry'[63].

Christ's transformation of what He received (IViii)

Clearly, though, the sense in which Jesus Himself understands His mission is both in terms of His unique identity as the Son of God and, as it were, as one who is taking up the inheritance of being taught that they were delivered from slavery as the 'Lord brought us out of Egypt with a mighty hand' (Dt. 6: 21). The visits that His family made to the Temple in Jerusalem, and indeed their whole work of transmitting the history of salvation to Jesus, are a part of that human assimilation of His religious inheritance that, in due course, would inform the life and indeed the liturgical action of Christ. In other words, there was a long process by which Christ assimilated what He was then to transform, in particular, the Passover (cf. Lk. 2: 41-42). As this is the last time that the Gospel mentions Joseph, except in passing (cf. e.g. Lk. 3: 23[64]), this identification of the Fatherhood of God by Jesus Himself, becomes very significant; it is as if Jesus had begun a process of differentiation which, in due course, would both identify who He is and, at the same time, increasingly confound His own family (cf. Mt. 12: 46-50; Jn 7: 1-9). In other words while Jesus Christ is going to bring an extraordinary

---

[63] http://www.jewfaq.org/barmitz.htm

[64] But note that there has been a revival of devotion to St. Joseph and that the Church has included him in the canon of the Mass after Mary (Pope John XXIII, quoted in *Redemptoris Custos,* 6) and, more recently, Pope Francis has extended the inclusion of St. Joseph's name in all the canons of the Mass.

dynamism to His public ministry and turn, as it were, as much "outwards" as the childhood years were "inward" and undisclosed, He is also going to transform His relationship to His mother and, ultimately, to all of us, such that we all share the "outreach" of God to us.

There is, then, a definite sense of Jesus fulfilling His human sonship in the remaining years in the family before His public ministry; and, as such, living out to the full the family as a place of formation, as a place of dwelling with the word, in Hebrew a *yeshiva* (meaning to dwell with the word[65]): an expression almost more applicable to His family than to Jesus Himself. Nevertheless, given Jesus' own period in the desert (cf. Mt. 4: 1-11), whereby He fulfilled Israel's dependence on providence (cf. Nm. 32: 13), we see that Jesus says of the word: "Man shall not live by bread alone, but by every word that proceeds from the mouth of God" (Mt. 4: 4). Thus the inner being of the Word made flesh is completely fulfilled in His dependence on 'every word that proceeds from the mouth of God'. Ultimately, Jesus resolves the question of His vocation in terms of a complete and perfect fulfilment of the *Shema*. When, therefore, Jesus is asked about the greatest commandment of the law, He is able to say: "You shall love the Lord your God with all your heart, and with all your soul, and with all your mind'" (Mt. 22: 36-37) and, going on to give the second commandment, He spoke of the love of 'neighbour as yourself' (Mt. 22: 39). In the end, however, all this is expressed on the cross: both obedience to God and love of us: 'as God formed the woman from the side of Adam, so also Christ has given us the water and the blood from his side in order to form the Church .... his Bride'[66].

---

[65] http://en.wikipedia.org/wiki/Yeshiva, more literally the word 'Yeshiva' means 'sitting'.

[66] A Reading from the "Catechesis" of St. John Chrysostom, bishop (Cat. 3: 13-19) p. 317 of *Readings for Lent and Paschal Time*, Vol. II, *pro manuscript: For private circulation only*.

Cornelius, the Roman Centurion, in the Acts of the Apostles (V)

We must also note, however, that whole families became Christian while, at the same time, having only gentile roots; thus, in the Acts of the Apostles, there is the notable case of Cornelius, a centurion and his family, who converted directly to Christianity (cf. Acts, 10); however, although this circumstance occurs early on in the spread of Christianity, it is not at all clear that it would be evidence of a different "influence" on the practice of marriage or the celebration of it.

Aside from the question of cultural influence, St. Paul draws a conclusion about a marriage between unbaptized persons: 'a lawful and consummated marriage between unbaptized persons can be dissolved when one of them converts to Christianity and the other opposes the faith or desires to be separated from the newly baptized spouse (7: 12-15 [of the *First Letter to the Corinthians*])'[67].

St. Paul and the Letter to the Ephesians[68] (VI)

In general, there are numerous aspects to the teaching of St. Paul on marriage and, as such, it can be implied in some general accounts that it was not regarded positively: 'The New Testament has a negative attitude to the sexual impulse and regards celibacy as a higher ideal than marriage (Matt.

---

[67] *Catholic Bible Dictionary*, General Editor: Scott Hahn, Corinthians, Letters to the: 1 Corinthians, p. 164.

[68] *Catholic Bible Dictionary*, General Editor: Scott Hahn, Ephesians, Letter to the, p. 246: Assuming Pauline authorship, unquestioned for 'seventeen centuries', this and other 'captivity Epistles' (e.g. Colossians) 'can be dated approximately to the early sixties'.

19: 10; I Cor. 7)'[69]; however, more precision is necessary as the New Testament teaching on marriage does involve a recognition of numerous aspects, particularly in the writings of St. Paul[70]. What emerges, however, given the Jewish emphasis on marriage and the traditions that developed and communicate its meaning, is that there were well established practices, liturgical rites and celebrations[71] which show us that Christian marriage *emerged out of a real and rich tradition* which, nevertheless, with the coming of Christ, was given a new beginning in what came to be understood as the sacrament of marriage.

What is particularly decisive in this "articulation" of the sacrament of marriage is the theology of St. Paul in the *Letter to the Ephesians*. By contrast, for example, in St. Paul's *Letter to the Colossians,* although there is some similarity of content (cf. Col. 3: 18-25) there is no mention of this new relationship between Christ and the Church and marriage (Eph. 5: 21-33). It is in the *Letter to the Ephesians* that St. Paul 'develops the theological theme found in the OT of connecting marriage with the covenant between Israel and the Lord. In the NT, this theme is transformed into the image of the Church and Christ as bride and groom: "As the church is subject to Christ, so let wives also be subject in everything to their husbands. Husbands, love your wives, as Christ loved the church and gave himself up for her" (Eph. 5: 24-25). Evoking the teaching of Genesis that the two shall become one flesh, Paul adds, "This mystery is a profound one, and I am saying that it refers to Christ and the church" (Eph. 5: 32) (CCC 1616)'[72]. In other words, given the background of St. Paul in the teaching and practices of Judaism, it is *completely coherent that it should be St. Paul who sees the new relationship between God and His people expressed in the uniqueness of the relationship of*

---

[69] *Encyclopaedia Judaica*, Marriage, p. 1027; and cf. Michael L. Satlow, *Jewish Marriage in Antiquity*, Chapter 1, p. 25.

[70] *Catholic Bible Dictionary*, General Editor: Scott Hahn, Marriage, p. 582.

[71] *Encyclopaedia Judaica*, Marriage, p. 1028.

[72] *Catholic Bible Dictionary*, General Editor: Scott Hahn, Marriage, p. 582.

Chapter One

Christ to the Church; and, in view of the traditional understanding of Jewish marriage recalling the covenant between God and His people on Sinai, that St. Paul grasps the new expression of the relationship between Christ and His people *as involving a new understanding of marriage*. Nevertheless, there is a kind of similarity between a sacramental understanding of the marriage and the Jewish perception that 'When a man cohabits with his wife in holiness, the *Shechinah* is with them (*Kitvey Rabbenu Mosheh ben Nachman*, Mosad ha-Rav Kook, Vol. II, p. 323)'[73]. Thus, in the end, Christ really does bring a new reality and, as it were, realises the "spiritual wealth" of the Jewish tradition from which Christianity springs.

In what follows we will note a few early testimonies to the reception of the sacrament of marriage.

Early testimonies from Tradition to the Nature of Marriage (VII)

As we emerge from the apostolic tradition and begin to traverse the patristic tradition, both of which are dynamically related in that the latter is constantly returning to the former as to its origin and original inspiration, we can discern more clearly the constancy of the *sacramental tradition of marriage*. It may be, for example, that this tradition shows itself in the very discrete way of signalling a difference in how marriage is lived by Christians and perceived by others; and, therefore, in the *Letter to Diognetus* we read that Christians marry but do not share their marriage bed or expose their infants to death[74].

---

[73] Goldberg, *The Jewish People: Their History and Their Religion*, p. 377.

[74] books.google.co.uk/books?id=STD9bY_EySkC&pg=PT89&lpg=PT89&dq=Christians+do+not+share+their+marriage+bed+nor+expose+their+infants&source=bl&ots=FMoMBQeT8Y&sig=-_NeAFkTchXhL4f6WOMHMD5RTdk&hl=en&sa=X&ei=o341U4vWCe2B7Qbr34

St. Ignatius of Antioch speaks more explicitly of the nature of Christian marriage, in his "Letter to Polycarp", (67 [5, 1]) when he says: 'Speak to my sisters that they love the Lord, and be content with their husbands in body and in soul. In like manner, exhort my brothers in the name of Jesus Christ to love their wives as the Lord loved the Church (Eph. 5:25)'. What is interesting, too, is that St. Ignatius encourages those who want to marry to discern their motive: 'It is proper for men and women who wish to marry to be united with the consent of the bishop, so that their marriage will be acceptable to the Lord, and not entered upon for the sake of lust'[75]. There are two particular reasons, if not three, for commenting on this reference to 'the consent of the bishop'. In the first place, there is an implicit biblical reference[76] to the Book of Tobit, where Tobias prays to take his wife Sarah for the right motive: 'And now, O Lord, I am not taking this sister of mine because of lust, but with sincerity' (8: 7). Secondly, the involvement of the bishop evinces a sense of the interrelationship between the mystery of the Church and the sacrament of marriage. Furthermore, there is an almost "institutional" expression of how, previously, a marriage arose through the mediation of fathers and other family relatives implicitly communicating, as it were, the providential action of God, whereas now that discernment of the vocation to marriage is "ordered" to the life of the Church. Although, in a way, the testimony of the Council of Trent is well removed, historically, from this period, what emerges in the language of Trent is an echo of the whole process of discernment and involvement of the Church and liturgical variety:

D4Cg&ved=0CDQQ6AEwAg#v=onepage&q=Christians%20do%20not%20share%20their%20marriage%20bed%20nor%20expose%20their%20infants&f=false

[75] William Jurgens, *The Faith of the Early Fathers*, Vol. 1, selected and translated by W. Jurgens, Collegeville, Minnesota: The Liturgical Press, 1970: p. 26, excerpt 67: Letter to Polycarp [*ca. A. D.* 110]: [5, 1].

[76] Cf. Pope Benedict's discussion of how Old Testament language can be both explicitly and implicitly drawn upon, p. 15 of *Jesus of Nazareth: The Infancy Narratives*.

'*I join you together in marriage, in the name of the Father and the Son and the Holy Spirit,* or use other words according to the accepted rite of each province'[77].

St. John Chrysostom (347-407AD) teaches on marriage and says the family is a '*Domestic Church*' and presents a fullness of family Christian life which, for most of us, remains a goal to be accomplished and reveals how fully involved in the practice of the faith the whole family is intended to be[78], which has been taken up by the Second Vatican Council's doctrine on the Church. What is remarkable, however, is that there is a similar expression and substance in Judaism, whereby it is said that 'marriage ... establishes ... the 'little sanctuary" of the home'[79]. In other words, even if the expression of the Christian Faith in the family was called a *Domestic Church*, there was a similar and almost identical expression in Judaism of the *little sanctuary*; and, therefore, we can see that the Holy Family was both a *little sanctuary* and a *Domestic Church: both living out the fullness of Judaism and beginning again in Christ.*

We conclude this brief review with Tertullian (c. 200 AD[80]), however, who deserves a mention for his reflections on marriage and, in the course of which he raises some questions about another aspect of the thought of St. Paul on the sacrament of marriage. For it is Tertullian who reflects on St.

---

[77] From the Canons on the reform of marriage, Chapter 1, ("Tametsi"), *Decrees of the Ecumenical Councils*, ed. N. Tanner, 2 volumes, (1990); and in two conversations with Fr. Norman Tanner, it was clear that these early councils treated of "problems" that had arisen and, as such, presupposed a continuous and actual tradition of action and liturgical practice (26-27 March, 2014).

[78] ancientfaith.com/specials/sVeritatisSplendor_lectures/st_john_chrysostom_and_married_life 5/11/11

[79] Goldberg, *The Jewish People: Their History and Their Religion*, p. 371.

[80] http://www.tertullian.org/readfirst.htm 29/10/11

Paul's words about those who '*marry in the Lord*'[81] (cf. 1 Cor. 7: 39). Clearly, however, there are numerous references in St. Paul's first letter to the Corinthians which pertain to marriage and which need exploring; for example, the verses on vocation (7: 7, 17) and what the Lord says about marriage 'that the wife should not separate from her husband' (7: 10) and also that St. Paul encourages a widow to marry 'whom she wishes, only in the Lord' (7: 39). The latter is taken by Tertullian to mean the widow is to marry 'a Christian'[82] and suggests that the widow is herself a Christian; and, by that, it could mean to include the significance that to marry another Christian is to enter the sacrament of marriage. For, subsequent to the passage quoted, Tertullian writes his famous praise of marriage, extolling 'the happiness of that marriage which the Church cements'[83], which again could be referring to the sacrament of marriage. Furthermore, Tertullian extols the whole mutual help of husband and wife, understood as a mutual help to lead the Christian life, which is taken to be such a "modern" understanding of marriage. Tertullian says, for example, 'Together they pray, together prostrate themselves, together perform their fasts; mutually teaching, mutually exhorting, mutually sustaining. Equally (are they) both (found) in the Church of God; equally at the banquet of God; equally in straits, in persecutions, in refreshments'[84].

In conclusion, while the evidence for the passage of the Jewish rite of marriage "into" the Christian sacrament of marriage, beginning with the marriage of Mary and Joseph, would seem to be *profoundly implicit* in the

---

[81] To His Wife, http://www.newadvent.org/fathers/0404.htm 28/10/11, Book II, Chapter 1.

[82] To His Wife, http://www.newadvent.org/fathers/0404.htm 29/10/11, Book II, Chapter 2.

[83] To His Wife, http://www.newadvent.org/fathers/0404.htm 28/10/11, Book II, Chapter 8.

[84] To His Wife, http://www.newadvent.org/fathers/0404.htm 28/10/11, Book II, Chapter 8.

life of the Church, it does seem to reveal a profound sensitivity to the very human nature of this development. In other words, the *Incarnation of the Son of God* entailed a kind of *principle of incarnation* in the very transformation of the Jewish cultural life into which He came and through which He was to express the new life He came to give us. In the end, however, the Holy Family is both a wonderful communication of the mystery of the Blessed Trinity (cf. *Familiaris Consortio,* 11; *Letter to Families,* 6), an amazing interpenetration of the human and divine reality[85] and the human origin, as it were, of the transition from a Jewish inheritance to the Christian sacrament of marriage and understanding of vocational celibacy. Thus, while this essay began with the perception of the relevance of the Jewish rite of marriage to the Christian sacrament of marriage, what has been unexpectedly pivotal in the whole discussion is the "embodied" reality of this "moment" of transformation in the mystery of the Holy Family. Finally, while considering what seemed to be a tangent, namely the vocation of Christ to celibacy, it becomes clear that it was through the Holy Family that Christ encountered the formation that was to express His vocation as a fulfilment of the *Shema: loving God and man to the end.* In other words, the *inner logic of the Incarnation* shows itself, ultimately, in the Word "being from" the communion of the Blessed Trinity and that of the Holy Family and constituting, with the Holy Spirit, the communion of the Church; and, therefore, *the language of the celibate bridegroom is the ultimate spiritual expression to which marriage, from the beginning, has orientated the eschatological hope of salvation.*

---

[85] Cf. Hans Urs von Balthasar, *Mary for Today,* Slough: St. Paul Publications, reprinted 1989, p. 35; and cf. Lk 1: 26-35.

Foreword to Chapter Two

Maria McFadden Maffucci

It is an honor to have this opportunity to share some personal thoughts inspired by this exploration of Marian doctrine. Although I am a lifelong Catholic who has always prayed to Mary, I often felt something lacking; I prayed more directly and felt more intense love for Jesus. This has changed recently, through the grace of a deeper connection to Mary, and a better understanding of her role as the Sharer of the Sufferings of Christ, an attribute explored in the chapter which follows.

The birth of James and Our Lady of Sorrows

In 1993, I was pregnant with our first child. As the due date of August 31st came and went, I thought about significant dates and how nice it would be for the baby to come on my best friend's birthday, September 4th, and then, September 8th, the Blessed Virgin's birthday! But that came and went as well. Finally, after two days of induced labor and a c-section, James Anthony arrived, on September 15th. I was disappointed to see that it was indeed a Mary feast day, but Our Lady of Sorrows. I didn't know why — here was a joyful event, balancing a great sadness in my family – my 34-year-old brother was dying of cancer. The connection of sorrow to birth was unsettling.

When James was four, he was diagnosed with autism, which was devastating at first, but then his diagnosis became just part of the wonderful child he was. We had two more children, and life had its challenges, but there was much joy. At times I thought about Our Lady of Sorrows and thought yes, well I do sort of understand now the significance of the date, but I still kind of kept Mary at arm's length.

In his young adulthood, however, James' disabilities have become a crisis, as he has had a mysterious regression. His situation and what to do about it have pushed us to our limits, and caused intense anxiety for me as his mother. Mothers are supposed to know how to make things better, and I can't!

"I started to experience great comfort in understanding her as Mother of Sorrows"

Partially from desperation, I embarked on a 54-day rosary novena to Mary. One is to pray a full rosary for 27 days for an intention and then immediately 27 more days in thanksgiving — no matter if prayers are answered, trusting that God does what is best. What I found as I prayed to Mary first thing every morning: I started to focus less on what I was begging for, for my son, and more on Mary herself, and I started to experience great comfort in understanding her as the Mother of Sorrows. Each day I brought my sorrow and worries to a mother who shared them in her mother's heart. I was able to surrender my attempts at control, and put things in the hands of the Lord, her beloved Son. Mary's role as Intercessor and Advocate became so much more real.

I have learned that when things go wrong, really wrong, the world offers no comfort, but we can go to Mary. By entering into the mystery of the passion of Jesus, with Mary at our side, somehow, we can bear the unbearable, because Mary represents to us the bridge between our human weakness and pain and the glory of Jesus Christ.

And, let us not forget, though Mary suffered the unspeakable agony of the loss of her child, three days later, she saw Him again! And if we mediate on that moment, in the midst of our distress, we have all the faith and hope that we are meant to have to sustain us. Love is eternal, and more powerful than death. With the knowledge of the resurrection we can persevere through suffering and know that as with Mary, God had a plan for all those who trust in Him.

It has taken me many years, but I now thank God that James was born on Our Lady of Sorrows. And I thank God that through this difficult time with my son He has offered me new hope through closeness to Mary.

A deeper embrace of "the Mystery of Mary"

I am not a theologian, but Etheredge's proposal in these pages for a new dogma regarding Mary is intriguing. For contemporary Catholics, it can be tempting to play down devotion to Mary in an attempt at ecumenical harmony – but what a mistake that is. Etheredge speaks about God's choice of Mary — a particular woman at a particular point of history. Mary wasn't just a "good women" who played an important part of God's plan; that's akin to saying Jesus was a "very good man" but not God, which C. S. Lewis refuted in his famous "trilemma" (either He was who He said He was, or he was a lunatic, or worse, a devil). Mary was chosen by God, before time, to be born without original sin, which elevates her above all other humans. Her perfect Yes to God in the face of the unknown, her acceptance of suffering and joy, is our model. This cannot be overstated.

Integral to Etheredge's proposition is that the right understanding of Mary's role also gives us an understanding of the essence of femininity, something sorely needed in current times. As a Catholic woman working in the prolife movement, I see more and more that at the root of the glorification of abortion is an evil attack not just on the vocation of motherhood but on the very nature of being a woman. Etheredge points to this when he writes: "Just as Mary is chosen as the Mother of the Lord, in view of the gift of spousal womanhood, so women are entrusted with the welcome of a child; indeed, it may well lie in the psychological makeup of the woman to grasp the wholeness of human being as a gift to be welcomed and nurtured."

Whether we are biological mothers, adoptive or foster mothers, or spiritual mothers, women when they follow their true nature have a unique capacity to receive others. Human beings have a deep longing for that kind of receptive love, even as some denigrate it through making motherhood, marriage and even biological reality a matter of personal "choice" and individual fulfilment.

If we embrace the Mystery of Mary more fully, as Etheredge here urges, perhaps in a new dogma, she can lead us deeper into the Truth of God's purpose in creating man and woman, and help us find peace and renewed hope in the midst of suffering.

Chapter Two

Part I of a Marian Triptych:
Mary is the Choice of God

General Introduction to Chapter Two: A Marian Triptych. Why call the next three chapters a "Marian Triptych"? On the one hand a triptych is normally a set of three paintings and, therefore, what presents itself is a threefold image of what is visible. What is visible almost immediately recalls the Son of God who became visible through the mystery of the *Incarnation* (cf. Jn. 1: 14); and, therefore, what is visible is what is invisible: the body manifesting the soul: man and woman showing forth the mystery of relationship: the human manifesting the divine. In other words the reference point is not the discussion, however interesting, but the reality of personhood: both human and divine; and, in this instance, what is to be considered is that Mary is the concrete choice of God. In a particular way, then, the concrete choice of God comes with an enduring significance in terms of the constant challenge to understand the threefold providential workings of God: in the relational reality of womanhood, in the dynamic mystery of Christ and His Church and in terms of the significance of the times in which we live.

Without, however, distinguishing any one of these it is probably significant for all three that Cardinal Ratzinger, now Pope Emeritus Benedict XVI, says: 'If the misery of contemporary man is his increasing disintegration into *mere* bios and mere rationality, Marian piety could work against this "decomposition" and help man to rediscover unity in the center, from the heart'[1]. On the one hand there is the vividly painful reality of what happens when 'bios' and 'rationality' are disintegrated. In the case of "using" the human embryonic child for experiments there is a kind of dichotomy between the child and "biological life"; and, therefore, "rationality" in this case is an almost active denial of the full humanity of the subject – for human conception is the start of a process of maturing the manifestation of the person conceived. It is as if the very fact that a human child can be manipulated in a glass dish changes the reality of what is in the dish; but, tragically, nothing has changed in terms of the child: the conception of a human person is an irreversible reality. On the other hand, 'Marian piety could work against this "decomposition" and help man to rediscover unity in the center, from the heart.' In front of the mystery of human conception, a mystery involving the action of God, a response from the welcoming depths of the woman's being is expressed in the words of Eve: 'I have gotten a man with the help of God' (Gn. 4: 1); and, even if the woman participates in the fall of creation as well as the man, so the experience of these sufferings is not without significance for the healing of the wounds of sin (cf. Gn. 3: 16-19). Just as Mary is chosen as the Mother of the Lord, in view of the gift of spousal womanhood, so women are entrusted with the welcome of a child; indeed, it may well lie in the psychological makeup of the woman to grasp the wholeness of human being as a gift to be welcomed and nurtured.

In a world, then, which "uses" the emotional argument to exploit people's sympathy in response to suffering and to advance an answer of death to life's difficulties, whether it be abortion, euthanasia or destructive embryo experimentation, Mary's experience of suffering as a participation in the suffering of Christ is also an indication of where we are to find the roots of an answer to the black holes which threaten to destroy us. At the same time, however, as there is a medicine of life which seeks to cure and to alleviate the suffering we experience there is, too, a resurrection of meaning that awaits

---

[1] Joseph Ratzinger, p. 36 of *Mary: The Church at the Source*, 2005.

those who persevere in praying through the problems of life.

Chapter Two: Introduction: Where do we need to begin with a reflection on Mary, spouse of St. Joseph[2] and Mother of the Lord?

On the one hand, in the times in which we live, it could be argued that we need to recover the variety and depth of the vocation of women and not, as it were, stereotypically "crush" the identity of women into a kind of "mould" of Mary; and, therefore, a meditation on Mary needs to recover the historical reality of women's oppression and persecution and how it has "moulded" or even misshaped the identity of women. Thus Cardinal Ratzinger, now Pope Emeritus Benedict XVI, observes three types of erroneous feminism: 'Exalt the woman better to oppress her'[3]; 'Mary's song [the Magnificat] becomes the motto of a certain theology that considers it its duty to advocate the overthrow of established social structures'[4]; and thirdly, 'Mary as the emancipated woman who, uninhibited and conscious of her destiny, confronts a culture dominated by men'[5].

On the other hand, what if the claim that the historical reality of Mary's marriage to St. Joseph is true and that her virginal motherhood, mystery that it is, is equally true; and that, therefore, there exists a starting point not of our choosing but, as it were, expressive of the will, choice and love of God? In other words we may have more difficulty in coming to the reality of Mary, precisely because she is *the choice of God*, than because she seems to express a "stereotypical" image of woman; for, in a word, Mary communicates a "word" of God which passes through all false perceptions of women, indeed through all varieties of legitimate and good vocations and gifts of women and

---

[2] Cf. St. John Paul II, *Redemptoris Custos*.

[3] A small book in which a *Commentary* by Hans Urs von Balthasar and an *Introduction* by Cardinal Ratzinger are published with *Redemptoris Mater* by Pope St. John Paul II, (San Francisco: Ignatius Press, 1988); from the "The Sign of the Woman: An Introduction to the Encyclical" by Cardinal Ratzinger, p. 9.

[4] *Ibid*, p. 10.

[5] *Ibid*.

brings us, simply, to a challenge of faith: that it has taken centuries to grasp, slowly but irrevocably, that Mary is a true "word" of salvation for men and women of all time (cf. *Dei Verbum*, 11[6]). Perhaps if there is a true appreciation of Mary in salvation history there will be a brighter recognition of the contribution of the Church to the development of women and the unfolding of their talents for the many splendored benefit of others.

There are three parts to this "Meditation" on Mary[7]. In this, the first of a three part triptych, there are five aspects to be considered. I will answer my own interest in the question (I) and then continue with the liturgical expression: Mary 'Gate of Heaven' (II). It then seems opportune to consider the structure of Pope St. John Paul II Apostolic Letter *On the Christian Meaning of Human Suffering*. His letter sets out in a particularly relevant way the doctrine of St. Paul on the vocation of the Christian to be *Sharers in the Suffering of Christ* and the application of this doctrine to Mary (III). A fourth section constitutes a reply to some objections to a further dogmatic statement on the mystery of Mary (IV). Thus the conclusion will address the possible benefit of a further elucidation of the salvific mystery of God's choice of Mary in salvation history (V).

I: The origin of my interest in whether or not there needs to be a new dogmatic statement on the mystery of Mary

I read with interest a series of articles in *The Tablet* on the theme of *The Place of Mary*[8]; each of them, for various reasons, rejected the benefit of a

---

[6] The *Dogmatic Constitution on Divine Revelation*, Second Vatican Council.

[7] The second of these Meditations on Mary is called: "Part II of a Marian Triptych: Our Hope in Mary"; and, finally, the third is called: "Part III of a Marian Triptych: Mary and Prayer"; all three articles now comprise Chapters Two to Four of this book.

[8] These articles will be specifically referenced, later on, but they were published in 1998, in English, in England.

further dogma on Mary. I therefore offer this piece by way of a contrast to these positions and I do so for the following three reasons. Firstly, a legitimate difference, charitably expressed, is often a condition of the development of a discussion. Secondly, whether or not there is a new Marian dogma this discussion is transforming what seemed to me to be an obscure and scarcely relevant question concerning Mary into one with increasing significance for us all. Thirdly, is the *choice of Mary by God*, any different from asking about any other aspect of the history of salvation; indeed, is it not about asking why God acted the way He did in choosing the Hebrew people? In other words, does the choice of Mary "participate" in the choice God made of the Hebrew people and the history of salvation which has been expressed in terms of their history? Exploring "Marian doctrine" is, therefore, exploring what God is saying in the lived history of salvation.

While, however, these initial reasons were a certain starting point to these meditations on Mary, it is nevertheless true that the reality of salvation history being expressed in the language of 'the Word [that] became flesh' (Jn. 1: 14) in the man Jesus Christ and the complementary reality of the 'woman' (Jn. 2: 4), His mother Mary, are also encouraging us to "revisit" the very mystery of the creation of man, male and female, in the beginning. Thus, it was while looking again at what is called *The Nuptial Mystery*[9], the mystery of the union of God and man, that I found two short texts of Hans Urs von Balthasar worked on each other like chemical ingredients ripe for a reaction.

II: Mary 'Gate of Heaven'

The first of Balthasar's thoughts was this: 'the Father's act of self-giving, with which he pours out his Son through all space and time of creation, is the definitive opening of the very trinitarian act in which the "persons" of God,

---

[9] This was the theme of a lecture by Bishop Angelo Scola on the 21 March '98 at the Oxford University Catholic Chaplaincy, organized by Stratford Caldecott of the *Centre for Faith and Culture*.

"relations", forms of absolute self-donation and loving flow'[10]. And the second was that this *definitive opening of the very trinitarian act* is hidden within the very mystery of Mary's conception of our Blessed Lord; he says: 'The angel announced to her not just the incarnation but fundamentally the entire mystery of the Trinity'[11].

Now the present task is to say that it was through thinking about the fact that the Son of God *chose* to come to us *through* the "Gate of Heaven", as Mary is called in *The Litany of the Blessed Virgin*, that Mary is by God's own choice *put in the position she is.* In other words if there is anything to this theological reflection on Mary then it is because it is about formulating what is in the first place *the word within the deed of God (Dei Verbum,* 2) which is both the *Annunciation* and the *Incarnation*: first the Annunciation and then the Incarnation. Thus the task of theology is to read the work of God; and, therefore, what comes first in all this is the fact that God chose Mary to be the "Gate of Heaven". So the question arises: Why did God put Mary in a position which will inevitably mean that in thinking about our salvation in Christ we will be confronted as it were with the 'scandal' of a human being, indeed a woman, right at the open heart of it? The Church of Christ[12] recognizes that there is a sense in which Mary is the 'cause' of our salvation. In the words of St. Irenaeus, quoted with approval by the fathers of the *Second Vatican Council* in the *Dogmatic Constitution on the Church* (56), the Church says of Mary: 'she "being obedient, became the cause of salvation for herself and for the whole human race"'.

Thus one cannot escape the thought of the *coincidence* between the

---

[10] Hans Urs von Balthasar, *Neue Klarstellungen* (New Elucidations), 67-70, page 146 of *The von Balthasar Reader,* edited by Medard Kehl and Werner Loser, translated by Robert J. Daly and Fred Lawrence, (Edinburgh: T & T Clark, 1982).

[11] Hans Urs von Balthasar, *Mary for Today,* translated by Robert Nowell, (Slough: St. Paul Publications, reprinted 1989), p. 35.

[12] Cf. *Dogmatic Constitution on the Church, Lumen Gentium,* of the Second Vatican Council, 8.

following facts of our time: Pope St. John Paul II called for a 'new feminism' in his encyclical letter on *The Gospel of Life* (99); the current onslaught against the *woman bearing a child* is a kind of culmination of discrimination and prejudice against *the poor of the Lord* (cf. Lk. 1: 46-55[13]). At the same time there has been a growing reflection on Mary, the *Mother of God* (*Theotokos*), as the *Immaculate Conception* (in 1854), conceived without original sin and full of grace and the dogma of her *Assumption* into heaven (in 1950), body and soul. Then as a kind of culmination of an appreciation of her identity as a type of the Church, *Lumen Gentium* says in the teaching of St. Ambrose: 'the Mother of God is a type of the Church in the order of faith, charity and perfect union with Christ'[14] (in 1964). Drawing on St. Augustine's doctrine as quoted in *Lumen Gentium*, the Church says of Mary, "she is clearly the mother of the members of Christ"[15], and so St. Paul VI promulgates her title as 'Mother of the Church'[16] (in 1967).

Thus it is very much in the spirit of the times in which we live that there is a development of doctrinal reflection on Mary; indeed, in a certain way, Mary has come to be recognized as expressing a kind of "totality" of human being: at once completely human and religious. In other words, Mary is being identified as a "living" expression of an anthropology of the whole, even completely redeemed human being. Thus Cardinal Ratzinger says of her: 'The 'biological' and the human are inseparable in the figure of Mary, just as

---

[13] Article 37, p. 121 of a small book in which a *Commentary* by Hans Urs von Balthasar and an *Introduction* by Cardinal Ratzinger are published with *Redemptoris Mater* by Pope St. John Paul II, (San Francisco: Ignatius Press, 1988).

[14] *Lumen Gentium*, 63, referring to the teaching of St. Ambrose: *Epist.* 63: *PL* 16, 1218.

[15] *Lumen Gentium*, 53: Cf. St. Augustine, *De S. Virginitate*, 6: *PL* 40, 399.

[16] *Signum Magnum*: http://w2.vatican.va/content/paul-vi/en/apost_exhortations/documents/hf_p-vi_exh_19670513_signum-magnum.html.

are the human and theological'[17]. In a very concrete way concerning the beginning of life, *Lumen Gentium* says that Mary is 'Enriched from the first instant of her conception with the splendor of an entirely unique holiness' (56). Thus if Mary was 'Enriched from the first instant of her conception', that implies that there is a personal presence from the 'first instant of ... conception' and, therefore, that just as Mary is conceived from 'the first instant' then we, expressing the same humanity, are conceived from the 'first instant of ... conception' too[18].

More widely, the Church recognizes a number of titles when she says of Mary: 'the Blessed Virgin is invoked in the Church under the titles of Advocate, Helper, Benefactress, and Mediatrix'[19]. Thus the particular title of 'Mary, Mediatrix of All Graces'[20], does seem to be characteristic of this renewed understanding of how fully and how intimately she participates in the redemption of the world, 'under and with' (*Lumen Gentium*, 56) her son Jesus Christ.

III: *Sharers in the Suffering of Christ*[21]

As a prelude to this section on suffering it is helpful to recall that St. John Paul II, like many of us, did not know how to react in front of the sufferings of others; he was, as we know, an exceptionally fit man who canoed, skid, swam, walked, cycled and endured long and exhausting journeys, vigils,

---

[17] "Thoughts on the Place of Marian Doctrine and Piety in Faith and Theology as a Whole", *Communio*, 30 (Spring 2003).

[18] Cf. too, Francis Etheredge, Chapter 12 of *Scripture: A Unique Word*, Newcastle upon Tyne : Cambridge Scholars Publishing, 2014: and Etheredge, *Conception: An Icon of the Beginning*, forthcoming from enroutebooksandmedia.

[19] *Lumen Gentium*, 62.

[20] "Mary, Mediatrix of All Graces", by Fr. William G. Most: https://www.ewtn.com/faith/teachings/marya4.htm.

[21] St. John Paul II, *Salvifici Doloris*: This is the title of chapter five.

negotiations and all kinds of difficulties. Between, then, his early days of health and his later sufferings because of parkinsonism, there was a period of learning about suffering. We too can take heart if we too are 'intimidated' by the sick. St. John Paul II says:

> 'I remember that at the beginning the sick intimidated me. I needed a lot of courage to stand before a sick person and enter, so to speak, into his physical and spiritual pain, not to betray discomfort, and to show at least a little loving compassion. Only later did I begin to grasp the profound meaning of the mystery of human suffering. In the weakness of the sick .... By their illness and suffering they call forth acts of mercy and create the possibility for accomplishing them'[22].

In keeping, then, with the fact that the context of all Christian doctrine is the mystery of the Blessed Trinity, it is worth adding a thought of St. John Paul II on the origin of our redemption in the heart of the Father. Two years after his *Letter on Suffering*, in his Encyclical Letter *On the Holy Spirit in the Life of the Church in the World*, St. John Paul II says that there is an origin to the mission of the Son of God in the reaction of God the Father to the sin of man. The Pope calls this reaction of God the Father, out of which comes the redemptive mission of the Son, His: *'fatherly "pain"'*[23]. On the one hand, then, there is the 'fatherly "pain"' of God the Father which is not expressed, as it were, abstractly – but in the suffering of the Son of God; and, on the other hand, there is the whole, multifaceted "face" of human suffering. Thus Pope St. John Paul II, as I have indicated, takes up the whole interconnected, almost feminine face to this sign of our times in the suffering of women amidst the multitude of sufferings with which the world abounds, and draws

---

[22] St. John Paul II, *Rise, Let Us Be On Our Way*, translated by Walter Ziemba, London: Jonathan Cape, 2004, p. 75.

[23] St. John Paul II, *Dominum et Vivificantem*, 39.

out of it a modern word on the theme fundamental to Christianity: What is the meaning of suffering?

After acknowledging the different kinds of suffering which correspond to the full reality of the human person, the Pope considered the theme raised in the Old Testament book of Job: *What mystery of God is hidden in the fact that God permits the suffering of the innocent?* This theme is relevant to our question concerning Mary because she is the one who remains faithful in her innocence[24] to the gift of her own *Immaculate Conception*: a gift which she receives in 'view of the merits of Jesus Christ', as it says in *The Definition* of this dogma[25].

The reflection on Job is a prelude to his meditation on the sufferings of Christ, the sinless one. It is the suffering of Christ which is the sole cause of our redemption (cf. 1 Tim. 2: 5): a suffering which is, as it were, in the relationship of the Father and the Son[26]; and it is a suffering which does, therefore, *intimately* involve the Holy Spirit[27].

In chapter five of this little masterpiece St. John Paul II goes on to show the development, in the writings of St. Paul, of the *Christian doctrine of the continuation of the sufferings of Christ in the sufferings of the Body of Christ, the Church*; for St. Paul says: "Now I rejoice in my sufferings for your sake, and in my flesh *I complete what is lacking in Christ's afflictions* for the sake of his body, that is, the Church" (Col. 1: 24[28]). At the beginning of chapter six of *The Gospel of Suffering* St. John Paul II considers how Mary *receives through her vocation as the Mother of God the gift of her* 'contribution to the redemption of all': a contribution which is as 'unrepeatable' as it is of 'an intensity which can hardly be imagined.... '. Thus the Pope comes to say of Mary: 'She truly has a special title to be able to claim that she "completes in

---

[24] Cf. Pope St. John Paul II, *Veritatis Splendor*, 120.

[25] Pius IX, *Ineffabilis Deus: The Definition*.

[26] *Salvifici Doloris*, 18.

[27] St. John Paul II, *Dominum et Vivificantem*, articles 39 and 41.

[28] Quoted in *Salvifici Doloris*, 24.

her flesh" - as already in her heart - "what is lacking in Christ's afflictions"'[29].

In a word, then, would it be inconsistent with this development of Marian reflection if the Church was to define Mary as *Mediatrix of all Graces*? Indeed, if all graces are ordered to or flow from Christ and His Church, then all graces are related to 'the Catholic faithful, others who believe in Christ, and finally all mankind, called by God's grace to salvation'[30]. If all can intercede for the grace of God to help each other, then Mary can especially intercede for us all; and, if Mary is a type of the Church, then Mary's intercession pre-eminently expresses God's willingness to draw man into the work of redemption. If Mary, then, is the *Gate of Heaven,* in virtue of her personal involvement in the Son of God's *Incarnation*, then it follows that God Himself has brought about Mary's graced, wholly human participation, in His work of salvation: a participation that is as unbounded as God's redemptively generous love.

IV: Some objections to this point of view

1. *Would a new dogma be an obstacle to ecumenism?* It is Hans Urs von Balthasar who says, in his *Commentary* on Pope St. John Paul II's encyclical on *The Mother of the Redeemer*: 'each denomination should first explore the depths of its own beliefs rather than try to reach out, for these depths may indeed provide the common ground to meet the other'[31]. In other words, does this *would be dogma* uncover, as it were, common ground between the different denominations? Indeed does it uncover differences which are proportionately clearer, easier to discuss, and in some way encompassed between what already unites the followers of Christ - *precisely because we are*

---

[29] *Salvifici Doloris,* 25.

[30] *Lumen Gentium,* 13, but cf. also 14; and cf. also *Gaudium et Spes,* 22.

[31] P. 162 of a small book in which a *Commentary* by Hans Urs von Balthasar and an *Introduction* by Cardinal Ratzinger are published with *Redemptoris Mater* by Pope St. John Paul II.

called to consider anew the very first thing, which is not what do we think but what has God done?

2. *Is this proposed Marian dogma Scriptural?* It has been shown, albeit all too briefly, that the doctrine of the Christian as co-redeemer 'under and with' Christ is pre-eminently true of Mary but also of each interceding member of the Body of Christ (cf. *Lumen Gentium*, 56) and that this is indeed the doctrine of St. Paul. The application of this doctrine to Mary in a way which clarifies what is "unique" to her contribution to our salvation which *God chose for her as Mother of our Savior*, is an application of a principle enunciated by Cardinal Ratzinger: that our understanding of Mary is first 'anticipated' in our understanding of the Church[32]. At the same time, just as the Church discovered, as it were, the doctrine of Mary's *Immaculate Conception* in the Scriptural evidence of the angel Gabriel addressing her as 'full of grace' (Lk. 1: 28), so it is possible that the Church is in the process of articulating the fullness of her identity as revealed by Christ on the cross when He said to His mother: "'Woman, behold your son!" Then he said to the disciple, "Behold, your mother!"' (Jn. 19: 26-27).

3. *In what sense can Mary be the Mediator of graces which were given before she even existed?*[33] Is this not a "part" of the same mystery by which the grace of our Redeemer was applied to Mary's own *Immaculate Conception*, prior to the fact of the coming of Christ? And, just as Christ's own redemptive work is efficacious throughout time, implying a relationship to everyone, from the beginning, so in a certain way His relationship to His Mother Mary is inseparably implicated in His redemptive mission of redeeming love. In other words the work of God does not follow the ordinary chronology of time; but, nevertheless, it is expressed in terms of the relationships which are integral to it.

---

[32] Cardinal Ratzinger, *Daughter Zion*, translated by John McDermott, SJ, (San Francisco: Ignatius Press, 1983), p. 68.

[33] Cf. *The Tablet*, 31.1.98, p. 153.

4. *Can Mary be the Mediator of sanctifying grace?*[34] This question directs us to the mystery that the Father first gave His Son to Mary *through* whom the Father gave His Son to us all. The possibility, then, of Mary being the Mediator of *all* grace *follows on* or is an inseparable part of her vocation, what Pope St. John Paul II calls her 'maternal mediation'[35]. In other words, and one has to continually come back to this: if God chose to come to man through a woman, then this woman is the way through which God gave us 'the grace and truth' which 'came through Jesus Christ' (Jn. 1: 17). Just, then, as the person of Jesus Christ is perfectly 'one in body and soul' (*Gaudium et Spes*, 14), being the bearer of the whole redemptive grace of God, so Mary is chosen by God to be the "Gate" through which we enter and receive all that God gives us; and, therefore, the whole of what God gives us which includes the fullness of His Trinitarian Being and the whole reality of redeemed relationships, beginning with our relationship to the Blessed Trinity, to His Mother and to all the redeemed. In other words, God having entrusted His very self to Mary, entrusts everything He is to her, too.

5. *Would such a dogma of Mary being Mediatrix of all Graces elevate Mary to* 'a status which is divine, not human'?[36] The vocation of the Christian, according to St. Peter (2 Pet. 1: 3-4) and the *Catechism of the Catholic Church*, involves the mystery which is expressed in the words of St. Athanasius: 'those in whom the Spirit dwells are divinized' (CCC, 1988). Thus it is a question of Mary being the created creature that the rest of us are *and the particular recipient* of the extraordinary favor of God (cf. Lk. 1: 28-30) which is yet a gift adapted to the one to whom it is given. On the one hand, then, the very humanity of Christ entered, inseparably, into the mystery of our redemption; and, as I have repeatedly said, all that is essentially human entered therein,

---

[34] Cf. *The Tablet*, 31.1.98, p. 153.

[35] *Redemptoris Mater*, 40, p. 130 of the previously cited little book; cf. also the first sentence of article 41, p. 131: 'her mediation, subordinate to that of the Redeemer ... '.

[36] *The Tablet*, 7.2.98, p. 185.

beginning with His relationship to His Mother. Is it for us, then, to reject the generosity of God in including everything that is fully human in His plan of redemption and, as it were, showing us its resplendent fullness in the very act of redemption? On the other hand, in the very mystery of the divinization of all truly human relationships, lies the mysterious nature of the Church and all her members: 'In order that we might be unceasingly renewed in him (cf. Eph. 4: 23), he has shared with us his Spirit who, being one and the same in head and members, gives life to, unifies and moves the whole body. Consequently, his work could be compared by the Fathers to the function that the principle of life, the soul, fulfils in the human body' (*Lumen Gentium*, 7).

6. *What is the justification of the doctrine that Mary can "dispense" the grace she receives?*[37] The answer to this riddle lies in the heart of Mary becoming, at the *Annunciation*, the 'cause' of her own and our salvation. The act of consent by which Mary becomes the Mother of God is the act of consent by which a mother brings a child into the world - *not exclusively for herself, but primarily for the sake of the child and, as it were, in anticipation of all the relationships into which that child will enter on entering human society*[38]. In other words the full human reality of the *Annunciation* is an act in which Mary "gives" what she herself has "received", just as a mother *gives* to the family of man the child she has first *received* (cf. Gn. 4: 1). In the relationship of Mary, therefore, to her son Jesus Christ, is the acceptance of the will of God that her motherhood pass, as it were, through the cross (cf. Lk. 2: 35); and, in the very nature of Mary's love being taken up into the redemptive love of her son, lies the mystery that her relationship to Him is according to the reality of motherhood (cf. Jn. 2: 1-12): that she exercises a

---

[37] Cf. *The Tablet*, 7.2.98, p. 185.

[38] Cf. p. 192 of an article by Antonio Sicari, *Mary, Peter and John: Figures of the Church*, (based on the work of Hans Urs von Balthasar), *Communio*, Vol. XIX, No 2, (*Summer* 1992).

ministry, as it were, which is a "redemptive transfiguration" of being the Mother of God and the Mother of the Church. In that the whole human reality entails freedom of action, is it not expressive of the "completeness" of Christ's redemptive love that the whole, living, free expression of Mary's being and action is wholly active in being Mother of Christ and Mother of the Church?

7. What about the implication of this doctrine that Mary is the Mediatrix of all Graces? *Is there a sense in which it would make all Christians "givers of the grace of Christ"?* Would it "endanger" the unique nature of Christ's salvific mission if His Church "communicated" His saving work to the men and women of each generation? Immediately there comes to mind the Church's giving of the sacraments of Christ. Is this part of the "'Marian' principle' which is an inseparable and complementary part of the Church of Christ?[39] In other words, God is not jealous of the treasury of His Son's death and resurrection; and, in so far as anyone is a disciple of the Son anyone can, through intercession, through the ministry and gifts they possess for the benefit of others and through the grace He gives to be at work in us, anyone can be a petitioner and indeed beggar of the gifts of Christ for those who need His help.

V: Conclusion

Now the question that surfaces for me at the end is the same question that surfaced for others: *What difference will it make if there is a dogmatic formulation of what are, at best, the doctrines of the Christian Faith concerning Mary?*

One possibility is that it will make our belief in the work of God which can be called the "mystery of Mary" so increasingly fundamental to our faith that the repudiation or rejection of it would be a repudiation and rejection

---

[39] St. John Paul II, *Letter of Pope John Paul II to Women*, 11.

of our salvation in Christ.

But why do this? A possible justification for such an action would be if it were to unequivocally lead people *to their salvation in Christ.* In other words it would be like saying that *God's own choice of Mary's contribution to our salvation* is a work of His to which He calls us; and he would call us to contemplate this particular work of His *because it will yield a fruit for our salvation which will be of particular relevance to the ongoing task of the Church.* Perhaps such a dogmatic formulation is a singular way of opposing the destruction of the dignity of the poor, particularly the dignity of child bearing women, by proclaiming the dignity of this exemplary Mother of the Lord: as if Mary is in some way able to mirror, through the gift of God, the irreproachable dignity of all the poor of the Lord.

Finally, perhaps the answer lies more fundamentally in the completeness of the redemptive love of God becoming known for what it is; and, therefore, it is not so much about extrapolating from what God has done and is doing so much as showing, as completely as possible, how completely redemptive the love of God is! Thus a more explicit recognition of Mary as *Mediatrix of all Grace* is about revealing how integrally complete is Christ's Redemptive Love of mankind, beginning with His Mother Mary, *the better to appreciate what becomes of us in the divinization of mankind which the love of God makes possible.*

FOREWORD TO CHAPTER THREE

LAURA ELM

In this beautiful and illuminating Foreword to Chapter 3, "Our Hope in Mary", Laura Elm, *Executive Director and Founder of Sacred Heart Guardians and Shelter*, really takes up the theme of *hope*. In this marvellous account of her experience she helps us to understand the "disconnect" between working in the *in-vitro* fertilization (IVF) industry and the humanity of the child at conception; and, at the same time, she brings the hope that just as it was possible that this "disconnect" was dissolved when Laura came to see the beginning of human life for what it is[1]: a true beginning of the human person, so it is possible that others will be helped too, from the most helpless of all to those rendering them so helpless.

Foreword to Chapter Three: Mary's Living Hope

Sarah, the first of my four children, was born on March 25th, the Feast of the Annunciation. At that time, I was very early in my Catholic faith

---

[1] Note the following experience: 'former abortionist Dr. Kathi Aultman says her journey began when the birth of her own child focused her mind on the humanity of preborn babies' ("Ex-abortionist: After giving birth, I made the 'fetal-baby connection'", by Calvin Freiburger:

https://www.lifesitenews.com/news/fetal-baby-connection-during-own-pregnancy-began-ex-abortionists-journey-to-life?utm_source=LifeSiteNews.com&utm_campaign=e44897702a-ProLife_11_12_2019&utm_medium=email&utm_term=0_12387f0e3e-e44897702a-401741553).

formation, and the significance of her birth date was lost on me. As I continue to grow in faith, learning more about God's love and our redemption, I am awestruck at this hopeful, joyful mystery.

I wonder what Mary thought about during her pregnancy. As she eagerly awaited her Son's birth, I imagine she tried to picture what He would look like and what kind of personality He would have. Maybe she pondered what she would be like as a mother, what it would be like to care for the child who was her Lord. Mary's heart surely overflowed with love, wanting only good things for the most precious life growing inside her. And all the while she readied herself to be the most loving mother, the most faithful disciple by quietly and obediently aligning her will to that of God.

Like so many, I desire to be more like Mary. And I hope God will put me on a path where that can be made possible. I also hope he will give me the graces to lovingly accept His path for me, and that I not be tempted to *fix* it when it becomes too difficult or *redirect* it because I think I can devise a smoother, happier way. I hope He will help me not get in the way of my own sanctification.

Awakening to the Truth

My husband and I hoped to have children, but both of us were shocked when God answered our prayers a little bit sooner than what we had in mind. We had been married not quite three weeks when, after a trip to an urgent care clinic for unusual abdominal pain, the doctor helped me understand what some of my lab results meant. Once we regained our senses, we were so excited.

In my mind, Sarah came into being on the day of that lab test. It would be several years before I learned how very busy she had been before I even knew she was in there. During those almost two weeks prior to the pregnancy test, hers had been the ultimate "inner life," growing and traveling in total silence and anonymity.

As a mom and a professional working on the business side of the infertility business, you'd think I would have more deeply digested the basics about prenatal development. But even though my days were spent analyzing data about IVF "success rates": retrievals, transfers, average number of embryos per transfer, percent of cycles started resulting in a live born "singleton," it just never occurred to me to think about what (or rather, who) these numbers where counting.

And then one day it did. Is an embryo a human being? Even when it's only one cell? A little research gave me a lot of proof that the answer is most certainly "yes." I wondered why no one in my profession ever offered training on the nature of the embryo. Perhaps it's so obvious that it goes without saying. But maybe we've been "without saying" for so long that the truth is no longer obvious. Or acceptable.

Based on how Assisted Reproductive Technologies (e.g., *in vitro* fertilization, or IVF) manufacture, manipulate and discard human beings, I knew I needed to find a new job, which was a punch to my pride on many levels. Thanks be to God that I was able to move to a different line of work.

But the experience also put something on my heart. I couldn't be the only one who didn't know how IVF harms the youngest human beings. Was there something I could do for our littlest brothers and sisters, even if it was just a small act of recognition?

Called to Witness

I founded Sacred Heart Guardians and Shelter (SHG) in 2017 with the mission to serve the smallest human beings: the human embryos whose lives start and end in Assisted Reproductive Technology laboratories. SHG reaches out to doctors and embryologists at fertility centers and offers to provide burial for the embryos who die while in their care.

By all accounts, this mission seems hopeless.

As of November 2019, SHG has provided burial for 120 deceased embryonic human beings. All the others – the hundreds of thousands of human beings who were created to die in the course of *in vitro* fertilization (IVF) in United States fertility centers – are thrown out as medical waste. As far as statistics go, SHG has accomplished next to nothing. But we are trying. Please pray we can do more.

The reactions of fertility center staff to our offer to provide burial range from annoyed amusement and curt politeness to outright anger. Every now and then there will be a sympathetic listener, who at first mistakenly perceives our offer to be a solution to the dilemma of so-called "leftover" embryos: a "Catholic disposition option" of sorts. They say how nice it is that we provide a way for patients to get closure.

Tension mounts when I remind the IVF staff that embryos – even those who are frozen – are unique, living, human beings; thawing them for any reason, even burial, is an act that kills them; SHG never encourages an action that would end the life of a living human being. I also remind him or her that embryos die in the lab for many other reasons. "Fresh" embryos die as a result of spontaneous self-arrest or because of a poor lab grade or genetic testing result. "Frozen" embryos can die in technical malfunctions or fail to survive the cryopreservation freeze or thaw process. At this point, the clinic staffer quickly brings the call to a close, usually passing me off to someone else: someone in the lab, a medical director, administration, or marketing. And if that person is finally reached, he or she may allow burial…if the patient requests it. After all, they will say, embryos are the patients' property. But they are also quick to say patients rarely ask for anything like this. According to the clinic, there is "no demand" for this kind of thing.

I imagine there wouldn't be demand if patients don't know that embryos (also known as fertilized eggs, pre-embryos, zygotes, morulas and blastocysts) are human beings. I also imagine it would take a miracle for the clinics to explain this basic biology to their patients.

Even among otherwise educated doctors and highly-skilled scientists engaged in IVF, I encounter wilful denial and ignorance about the embryos' humanity. Is it self-preservation or poor training that leads them to incorrectly state that embryos are not really human beings, that they are not really alive? They may ask if we also bury eggs, and I remind them that eggs are reproductive cells whereas embryos are human beings. They may agree to the biology, but not to the moral treatment, attempting to differentiate between human being and person.

Is there a Success Rate for the Embryo?

IVF continues to gain social acceptance and utilization. More and more, society celebrates those who share their "brave" use of IVF. IVF is no longer limited to couples with diagnosed clinical infertility, but also provides a way for those with "social" infertility (same sex couples, single people) to "build their family." As if a family can be purchased and manufactured, like a back porch or a garage stall. New laws in the US mandate insurance coverage and access to IVF. "Advancements" in IVF increase clinical "success rates," with more women becoming pregnant and giving birth to a live born baby.

But what about "success rates" for the embryo? The younger human being also monitored at the IVF clinic? Advancements for the patient's outcome (pregnancy) mean increased mortality for the embryo. Fresh embryos are kept in their culture media longer. After fertilization (i.e., a new human being has come into existence), approximately 50% of embryos die between culture day 3 (the previous standard day of transfer) and a Day 5 transfer (the new norm....some being in culture longer). Doctors increasingly strive to transfer a single embryo to the patient, avoiding the risk of multiple gestations and the ensuing health and financial liabilities. But that single embryo is carefully selected from among a sometimes large pool of his or her embryonic brothers and sisters, not only by visual grading, but also with an increasing use of Preimplantation Genetic Screening (PGS) and Preimplantation

Genetic Diagnosis (PGD). These methods help ensure that only the highest "quality" embryos (or those of the desired sex) are selected for transfer, or frozen for future "use". Freeze-all cycles are increasingly common: the patient has higher success rates achieving a pregnancy if the hormone protocol clears her system, but now *all* embryos are subject to the risk of freeze, thaw, abandonment or discard ("discard" is an industry term which means taking a living embryo out of either the incubator or freezer, bringing it to room temperature, watching until cellular activity ceases (it dies), and throwing it out). Donor eggs show higher patient success rates in some age groups, and these embryonic humans are intentionally deprived of the right to be born of and raised by their biological parent.

Our Call to Hope Against Hope

I hope against hope that IVF clinics will allow us to bury their deceased embryos. I hope against hope that no human being is treated as waste. The outlook is bleak as pro-abortion groups seek to overturn laws requiring miscarried and aborted fetuses to be buried or cremated. The cost of such unnecessary actions, they say, restrict a woman's "right" to an abortion. And these are human beings who look like us, with skin and bone, who can suck their thumbs and hiccup – Who can perhaps even survive *outside* the womb. If these human beings can be denied respectful and dignified treatment at their death, there is little hope for the human beings who sadly really do look like "a clump of cells."

More than anything, I hope against hope that each receptionist, nurse, doctor, lab technician, and embryologist will turn away from their IVF profession. I hope they will come to love God, in their heart and through their actions. I hope that I, spiritually weak and a chronic sinner, can know and do God's will. I hope Mary will pray for all of us, and that we can imitate her perfect love and faithfulness to God.

Mary, Queen of Heaven and Earth, pray for us.

CHAPTER THREE

PART II OF A MARIAN TRIPTYCH: OUR HOPE IN MARY

General Introduction to Chapter Three: Hope. 'It is on the path shown by this ... [sign of the woman] that we follow the trail of hope toward Christ, who guides the ways of history through this sign that points the way'[1]; and, more specifically, 'the originality of Mary's role of mediation consists in its maternal character, which aligns it with Christ's being born ever anew in the world'[2]. In other words, in a world in which motherhood is disfigured the motherhood of Mary is a sign of hope in the ever anew coming of Christ. But what is this hope?

As St. Paul says, 'faith, hope, love abide, these three; but the greatest of these is love' (1 Cor. 13: 13). Faith, hope and love are 'infused by God ... to make [the faithful] ... capable of acting as his children and of meriting eternal life' (CCC, 1813). By faith 'we believe in God and believe all that he has said and revealed to us, and that Holy Church proposes for our belief, because he is truth itself' (CCC, 1814). By hope 'we desire the kingdom of heaven and eternal life as our happiness, placing our truth in Christ's promises and relying not on our own strength, but on the help of the grace of the Holy Spirit' (CCC, 1817). By love or charity 'we love God above all things for his own sake, and our neighbour as ourselves for the love of God' (CCC, 1822).

---

[1] Joseph Ratzinger, p. 53 of *Mary: The Church at the Source*, 2005.
[2] Joseph Ratzinger, p. 55 of *Mary: The Church at the Source*, 2005.

The axis of eternal life, like the axes of human development, is to make possible the daily living in which all our joys and sorrows are experienced. Thus there is no contradiction between daily life and its demands and how 'the fruits of our nature and our enterprise ... [are] cleansed ... from the stain of sin, illuminated and transfigured, when Christ presents to his Father an eternal and universal kingdom' of 'justice, love and peace' (*Gaudium et Spes*, 39). In other words there is every reason to pursue the goal of alleviating, if possible, those suffering from infertility, debilitating diseases or the alleviation of the experience of the sufferings that can turn people towards the possibility of suicide. Respect for life, however, challenges us to recognize that there is an ethical framework within we work and which guides, practically, the development of bioethics; and, therefore, research into infertility is guided by seeking a way of helping which expresses the reality of the unitive and procreative significance of spousal love (*Humanae Vitae*, 12) and the sacredness of human life (*Humanae Vitae*, 13).

In terms of the specific theme of bioethics, then, we need to hope in the help of God in terms of the very medical procedures which are in accordance with His will; as, indeed, there is no certainty as to whether or not an illness can be cured, infertility remedied, or suffering illuminated and alleviated. In other words there is a possibility of passing through darkness, at times, in the course of persevering with our difficulties; indeed, 'For Mary, as for Abraham, faith is trust in, and obedience to, God, even when he leads her through darkness'[3]. Even in the case of persevering in the pursuit of an ethically upright way of helping those who have been frozen, as human embryos, there is a kind of "ethical darkness" until it becomes clear that a way can or cannot be found to help them; and, indeed, what are we to say of those who have been frozen in this unspeakably tragic disregard for the ongoing unfolding of their lives? Prayer and perseverance, then, are essential in the hope-led search to help those who suffer; and, indeed, like the wise

---

[3] Joseph Ratzinger, p. 49 of *Mary: The Church at the Source*, 2005.

men of the east, we need to search every aspect of human wisdom as well as appeal, constantly, to God to guide us.

Chapter Three: Introduction[4]. Just as the actual history of sin brought a problem between the sexes (cf. Gn. 3: 16-17) so the reality of salvation history entails a remedy for the problem between the sexes. In other words Mary, as the *Mother of the Redeemer*, both brings Jesus Christ into the world and, through Him, contributes to the restoration of the good order between the sexes (cf. Jn. 19: 26-27) which sin disfigured. Thus our relationship to Mary is an integral part of conversion to the reality of salvation history; and, therefore, our relationship to Mary is fundamental to recognizing the *order of salvation established by Christ*. The saving work of God in the life of Mary, the Mother of the Lord, is not only about how she intimately participates in the saving work of her son and the transformation of marriage into a sacrament of salvation[5]; it is, too, about how she participates in the plan of God as a whole: the salvation of the human race and, therefore, the reconciliation of man, male and female.

In the focus on hope of this second part of an essay on Mary[6], then, it is more than fitting that we hope in Mary; indeed, to hope in Mary is to receive the very prescription of our *Redemption in Christ*. It is God who has ordained the nature and history of our salvation. Therefore, pondering the mystery of what God has actually done, helps us to recognize how *perfectly salvation answers the reality of sin*. This essay considers the following question: What

---

[4] This piece first appeared as Part III of Chapter 9 of Volume III of a trilogy called *From Truth and truth: Volume III-Faith is Married Reason*, Newcastle upon Tyne: Cambridge Scholars Publishing, 2016, pp. 57-66.

[5] Cf. Francis Etheredge, https://www.hprweb.com/2014/08/the-holy-family-celibacy-and-marriage-a-reflection-on-the-passage-from-the-jewish-rite-of-marriage-to-the-sacrament-of-marriage/

[6] The first part of this essay on Mary was called: Part I: Mary is the Choice of God.

is hope? Therefore let us seek God *through* the very same 'Gate of Heaven'[7] that God Himself chose in order to seek us. In the first section of this essay we define hope in the reality of the Holy Family (I); the definition of hope in the light of the coming of Christ (II); then how many hopes appear to us as impossible (III); and a reflection on how Mary helps us to hope (IV); two objections to hoping in Mary (V); and, finally, a conclusion.

I: Hope in the Reality of the Holy Family

The history of salvation is, therefore, an actual remedy of the whole problem wrought by sin: a remedy that both addresses and transcends the real nature of our condition; and so marriage experiences a renewal which, it is possible to argue, is begotten in the mystery of the Holy Family's living the fulfilment of marriage as a sign of Christ's love for the Church: a sign expressed in the reciprocal love of St. Joseph and Mary which was, inseparably, open to embracing and rejoicing in the coming of Jesus Christ[8].

Thus Mary, spouse of St. Joseph, lives the redemption of marriage in the new beginning which arises out of their openness to Christ: Mary as His fruitful, virginal mother and St. Joseph as His celibate, virginal father. At the same time, however, they live this redemptive love in the reality of Mary as fruitful mother and Joseph as a "childless" father. Moreover, there is a hidden generosity in their act of faith in God which only comes to light, as it were, incidentally, although it could be said to be implied in the very "openness to God" which their openness to Christ expressed. Thus, it would seem, Mary and Joseph's "open house" showed itself in the extended family that was so closely identified with Jesus: 'Then his mother and his brethren came to him, but they could not reach him for the crowd' (Lk 8: 19); and, extending this,

---

[7] From *The Litany of the Blessed Virgin*, p. 15 of *A Simple Prayer Book*.

[8] Cf. Francis Etheredge, https://www.hprweb.com/2014/08/the-holy-family-celibacy-and-marriage-a-reflection-on-the-passage-from-the-jewish-rite-of-marriage-to-the-sacrament-of-marriage/.

Jesus replies: "My mother and my brethren are those who hear the word of God and do it" (Mt. 8: 21). The unbounded generosity of God finds expression in an increasingly unbounded, vocational relationship of Mary and Joseph to those around them; and, more widely, of their Son, Jesus Christ, to us all.

More particularly, in encompassing the whole human reality of husbands and wives being both childless and having children, there is a "sign" of this in the will of God for Mary and Joseph; indeed, a tender sign in that, when they lose Jesus, Mary affirms St. Joseph's fulfilment of the relationship of father to the child Jesus when she says: "Behold, your father and I have been looking for you anxiously" (Lk. 2: 48). Mary's affirmation of Joseph is all the more positive in view of how clear she is that Jesus is conceived of 'The Holy Spirit' (Lk. 1: 35, but also 26-56). On the one hand, lest there be any confusion about Jesus' identity His reply, when He is found, establishes a very clear distinction between the divine Fatherhood of God and that of Joseph: "Did you not know that I must be in my Father's house?" On the other hand, Jesus shows that He accepts wholeheartedly the fatherhood of Joseph as St. Luke writes, presumably drawing on Mary's testimony: 'And he went down with them and came to Nazareth, and was obedient to them' (Lk. 2: 51).

The hope, then, expressed in the Holy Family is that there is a "mission" for each of us – even if it looks, at times, as if God has taken what we thought it was from us; but, then, in the light of the will of God, we discover our vocation a-new in a way we did not expect.

II. The Definition of Hope and the coming of Christ

The hope that God gives us as a task[9] is the Christian hope of the *definitive coming of the kingdom of God*[10]: a kingdom which comes, *in a way*, through

---

[9] Cf. St. John Paul II, *Gift and Mystery*, p. 79.

[10] St. John Paul II, *Tertio Millennio Adveniente*, 46.

our 'concern for people in their concrete personal situations'[11]. This is because *of the intimate relationship between* what we do and to whom we do it[12]; and the fact that the coming of the kingdom is the coming of a person, the person of our Lord Jesus Christ: who has come; who is coming; and who is to come. Furthermore, the coming of a kingdom is the coming of a person because "person", means "relationship"; and, therefore, there cannot be the coming of Jesus Christ without the coming of all the relationships which, together, constitute the kingdom of God.

Our Christian hope, however, has the particular characteristic of *turning us to God for what we cannot do for ourselves.* The Christian 'must hope that God will give him the capacity to love Him in return and to act in conformity with the commandments of charity' (CCC, 2090). Thus *hope* is ordered to love. While hope will come to an end, *the end of love* to which hope is ordered, will not (cf. 1 Cor. 13: 13 and Rom. 8: 24-25); and so hope could be said to be ordered to the Holy Spirit *as to the Love*[13] *who will help us to love.* Furthermore, if the Holy Spirit is the "Person-Love" between the Father and the Son[14], *then we love God in our neighbour*[15] *in the "Love" that the Holy Spirit is*[16].

When St. John Paul II wrote about the second year of preparation for the beginning of the *Third Millennium*, he spoke of these things in connection with Mary as 'a woman of hope who, like Abraham, accepted God's will

---

[11] Cf. St. John Paul II, *Sollicitudo Rei Socialis,* 48.

[12] Cf. *The Holy Bible,* Revised Standard Version, Catholic Edition, San Francisco: Ignatius Press, 1966: Mt 25: 31-45. All Scriptural references are to this edition unless otherwise stated.

[13] Cf. St. John Paul II, *Dominum et Vivificantem,* 10.

[14] *Dominum et Vivificantem,* 10.

[15] *The Sacred History of Love,* an article by Antonio Sicari in *Communio,* Vol. XXIV, No 1, (Spring 1997), p. 19.

[16] Cf. St. Augustine's doctrine of the "appropriate attributions" of the Blessed Trinity.

"hoping against hope" (cf. *Rom.* 4: 18)'[17]. In other words, it is as if the *definition of Christian hope* is that it is a hope in God for what *is not humanly possible for man to do*. What emerges is that *Christian hope* is the *gift to man* which corresponds *to the need of man*. Therefore, when Abraham believed in God's promise of an heir, even though he and his wife knew 'it had ceased to be with Sarah after the manner of women' (Gn. 18: 11), Abraham was "hoping against hope" in the promise of the Lord. Thus, as Cardinal Ratzinger says: 'Now it becomes clear why barrenness is the condition of fruitfulness - the mystery of the Old Testament mothers becomes transparent with Mary'[18]. For without the *condition* of what is humanly impossible, *the reality* that "with God nothing will be impossible" (Lk. 1: 37) is not made manifest.

III: How many hopes appear to us as impossible?!

There seem to be a growing number of hopes that are like Abraham's "hoping against hope", in that the obstacles to a human accomplishment of these goals are too great, too numerous and too entrenched.

"Hoping against hope" for a conversion to reality, as Cardinal Scola once called it: the human reality of man, male and female, marriage and family life; and, "hoping against hope" for a recognition of the reality of sin, both in the Church and in the history of the human race, our relationships and society as a whole. "Hoping against hope", then, for a renewal of the Church as expressing the salvation of the sinner and a renewed appreciation of a philosophy that begins, as it were, with the everyday reality of our relationship to what exists.

The impossible hope of an end to an extremism that leads to a destruction

---

[17] *Tertio Millennio Adveniente*, articles 48 and 44-47.

[18] Cardinal Ratzinger, *Daughter Zion*, translated by John M. McDermott, SJ, San Francisco: Ignatius Press, 1983, pp. 52 and 70.

of people's lives, families, homes, Churches and the ordinary, everyday activities of earning a living, going to school, living in society and worshipping God. "Hoping against hope" for the help we need to abandon the arms race and the proliferation of nuclear weapons. The impossible hope for a dialogue between peoples about their past, the need for forgiveness and a common will to build a future together. To hope for the impossible hope of "Hoping against hope" for the return of kidnapped children.

To hope for the impossible hope for there to be an end to the misuse of power. Hoping, therefore, for political objectivity and a recognition of the real needs of the vulnerable peoples of the earth. To hope, then, for help to get to those who need it. "Hoping against hope" for a just and generous international world authority, a renewed concern for the working conditions of the peoples of the earth and the welcome of displaced people.

Hoping for an end for the technological exploitation of the human being and the environment of the human family; indeed, hoping for a decisive pursuit of the good of the peoples of the earth. Hoping, therefore, for a common recognition of the moral limits of technology, for the real promotion of the good of the human family and for the "gardening" of the earth.

Hoping for an end to the multi-national manipulation of peoples and profits; and, therefore, the widespread deprivation of reasonable hours, family wages and good working conditions. Hoping, therefore, for international wage agreements, profit sharing and benefits distributed to communities in need.

Hoping for an end to the misuse of human life from conception until death; and, indeed, all the multiple ways that people are enslaved and deprived of human dignity and human rights. Hoping, then, for a growth in the world-wide recognition of foundational human rights. Hoping, therefore, for the powerful to recognize the humanity of the weak.

How many different human hopes have turned out to be hopeless? How necessary it is to "hope against hope"; and, therefore, to hope in God for what it is humanly impossible to accomplish.

Either we are approaching the "end-times" or we are discovering, globally, that hoping in God is indispensable for the future of humanity – or both!

IV: How Does Mary Help us to Hope?

There are three complementary ways in which one can answer this question. On the one hand, we can believe that Mary *is a God given* 'sign of sure hope'[19]. On the other hand, we can ask for the help that Mary can give us because she is free to distribute the gifts of her Son[20]. It could even be said that the life of the Blessed Virgin Mary is an inseparable expression of both these elements: almost as if her life is an *embodiment of the gift of redemption*; for what was first effective for her and in her is what will be effective for us and in us. Thus there is what I will call the God given *ontological* reason for the fact that Mary is a 'sign' of sure hope; and then, on this foundation, there is the *unfolding* of her vocation within her own life and the life of the Church. For it seems as if the identity of Mary is like a glorious secret within the life of the Church: a glorious secret which the development of the Church makes it increasingly possible to communicate; indeed, as Cardinal Ratzinger, now Pope Emeritus Benedict XVI: 'the doctrine of the *Immaculata*, like the whole of later Mariology, is first anticipated as ecclesiology'[21].

---

[19] St. John Paul II, *Redemptoris Mater,* promulgated 1987, San Francisco: Ignatius Press, 1988, article 11, p. 64. This slim volume also contains an *Introduction* by Cardinal Ratzinger and a *Commentary* by Hans Urs von Balthasar. The references in this article are either to page numbers in the book as a whole or to the paragraph numbers characteristic of papal writing.

[20] Cf. *Redemptoris Mater*, article 21, p. 88.

[21] Cf. Ratzinger, *Daughter Zion*, p. 67; and cf. also p. 68.

What does it mean to speak of Mary's ontological identity?

This concerns the "gift" of what Mary is from the first instant of her creation. On the one hand Mary is a creature. She had a human mother and a human father and she came into existence at the moment of her natural conception[22]. And so she was a true daughter of the human race, born of and into the people of God. But on the other hand it is also and simultaneously true that *at the first instant of her conception* Mary was 'preserved free from the stain of original sin'[23]. Thus, just as Mary was conceived a true daughter of her parents she was made a true daughter of God[24]. In other words, Mary's *Immaculate Conception* is the gift of God; and it is the gift of God 'in view of the merits of Jesus Christ, the Savior of the human race'[25]. Furthermore Mary is given to be so splendid that in the extraordinary words of Pope Pius IX: 'she approaches as near to God himself as is possible for a created being'[26]. *Mary* is, therefore, *God's perfect creation*; after the Redeemer, she is the first *in the order of the redeemed*. And so it can be said that Mary is the "gift" to the human race of the beginning of the *new creation in Christ*: she is both the 'new Eve'[27] and 'the mother of the living'[28].

Thus Mary '*is the fulfilment of the promise God* made to man *after original sin*'[29]: the promise of a woman whom God destined to be the Mother

---

[22] Pius IX, *Ineffabilis Deus*, promulgated in 1854, Boston: St. Paul Books & Media, page 21: *The Definition*.

[23] Pius IX, *Ineffabilis Deus*: second paragraph: *Supreme Reason for the Privilege: The Divine Maternity*.

[24] Cf. Ratzinger, *Daughter Zion*, p. 70.

[25] Pius IX, *Ineffabilis Deus*, p. 21: *The Definition*.

[26] Pius IX, *Ineffabilis Deus*, p. 15: *Mary Compared with Eve*.

[27] *Lumen Gentium*, 63.

[28] *Lumen Gentium*, 56.

[29] I have adapted a phrase that St. John Paul II applies to the Incarnation in *Redemptoris Mater*, 11, pp. 62-63.

of His Son (cf. Gn. 3: 15). Thus, in the full mystery of her free acceptance of this, 'Mary remains a sign of sure hope' because she is a sign that God's election of her 'is more powerful than any experience of evil and of sin, than all that "enmity" which marks the history of man'[30]. In other words, Mary is a 'sign' of hope in the very being God gave her to be *precisely because* she *manifests,* as it were, God's faithfulness to Himself. God is a God whose acts[31] both reveal Him and ourselves to us. God fulfils His promise of salvation *from the seed of the woman*[32].

This identity of Mary is, however, both a gift and a task

Thus Mary is given a vocation at the very moment of coming into existence; but it is also a vocation which she is called to discern in the words and deeds of God. Moreover, the words and deeds of God exist in the life of the people of God: revealing the personal nature of the history of salvation. On the one hand, as the Mother of God, she is the most exalted of the people of God; and, at the same time, the most humble[33]: the most "totally dependent upon God and completely directed toward him"[34]. Furthermore, it is a vocation Mary is called to live out in the response and acts[35] of her own life. Thus in the famous words of St. Irenaeus, quoted in the last part of the Second Vatican Council's *Dogmatic Constitution on the Church*, which is devoted to Mary: 'she "being obedient, became the cause of salvation for herself and for the whole human race"'[36]: an obedience which St. John Paul

---

[30] *Redemptoris Mater*, 11, p. 64. I have reordered the last two sentences of this paragraph.

[31] Ratzinger, *Daughter Zion*, p. 61.

[32] *Redemptoris Mater*, 11, p. 63; and cf. Gn 3: 15.

[33] Cf. *Redemptoris Mater*, 35, p. 117 and 37, pp. 121-122.

[34] *Redemptoris Mater*, 37, p. 122.

[35] Cf. St. John Paul II, *Veritatis Splendor*, 71.

[36] *Lumen Gentium*, 56.

II calls an 'obedience of faith'[37]. In other words, the Annunciation was a *moment* in which God revealed to Mary the nature of the identity that was His gift to her; and in that same dialogue with the angel, God made that same 'gift' the subject of *Mary's free consent*.

Out of this *obedience of faith* by which Mary expressed her acceptance of what God had given her to be, there broke forth the "hope against hope" that "with God nothing will be impossible" (Lk 1: 37); and that *through her virginal consent* would come the saviour of the *family of God*[38]: the Son whom she loved in her heart before she conceived him in her flesh[39]. It was this "hope against hope" which, while conceived in the history of Israel, was in a pre-eminent way expressed in her own life. Thus, not only was this vocation from her own conception, but then it was also from the annunciation: from the conception of Christ. In the context of the people of God and her own life, there was the "hope against hope" of salvation *through the crucifixion of her Son* and what looked like the *complete 'negation'*[40] *of the promise of a redeemer*. Thus Mary "lived" the resurrection of her Son, His ascension, the birth of the Church and the transformation (cf. 1 Cor. 15: 51-56) of her own bodily self *in the mystery of her own assumption into heaven*.

Mary helps us to hope because God prevailed in her and, therefore, she embodied the fulfilment of the promise "planted" in her life; and, at the same time, she helps us to hope because, as our Mother, she "hopes in God" for us.

---

[37] *Redemptoris Mater*, 13, p. 66-69.

[38] Cf. John Paul II: a homily at *Belo Horizonte*, 1. 7. 80., on p. 126 of *Through the Year with Pope John Paul II*, edited by Tony Castle, London: Hodder & Stoughton, 1995.

[39] Cf. *Redemptoris Mater*, 13, p. 68.

[40] *Redemptoris Mater*, 18, p. 78.

V: Two Objections to our Hope in Mary

The first objection could be that as we are not immaculately conceived, then how can what befell Mary be of any use to us?

The answer is that the sacrament of Baptism is *our conception in Christ*. In other words, while there are important differences between the *Immaculate Conception* and our *Baptism*, there is a fundamental identity: each is an act of the Holy Spirit by which we are conceived in Christ; and each conception in Christ constitutes, as it were, the seed of a vocation to be discerned *through* the Church, among the people of God in whom the Father's acts of salvation are also and inseparably a word[41].

A second objection is that if Mary did not sin personally, then how can she know what it is for us to be sinners?

An answer to this is that *innocence* is no obstacle to experience[42]; for 'Not having known sin, she is able to have compassion on every kind of weakness'[43]. This is indeed a fascinating reversal of the popular idea that unless a person experiences something "he" cannot "know" it; however, what emerges from this "reversal" of that idea is the deeper, more mysterious wisdom of the cross. Thus the terrible nature of sin was experienced fully in

---

[41] Cf. *Dei Verbum*, 2.

[42] This question has depths which it is not possible to go into here; for instance, the "innocence of God" is an unfathomable mystery. But, put simply, it pertains to beauty: the resplendent beauty of a nature that is what it is intended to be; and, being perfectly true to itself, is resoundingly wonderful and unalloyed by any kind of taint, corruption or imperfection: a marvel to the eyes of faith. In other words, contrary to the lie of the devil (cf. Gn. 3: 5), there is a more profound knowledge than that of the experience of evil, namely, the balanced insight of innocence.

[43] St. John Paul II, *Veritatis Splendor*, 120.

the crucifixion, not because Christ personally committed sin, but because He fully experienced the sin-caused suffering of the crucifixion. Mary, therefore, sharing in the suffering of her son, "understood" the "disfiguring reality" of sin through what it did to her son.

Mary, then, not only knew "sin" through the crucifixion of her son, but she also knew the forgiveness of sin through the "paschal" coming of the Holy Spirit; and, therefore, it can be said that Mary's knowledge of salvation was thus as humanly complete as it was possible to be. In other words, although she had not sinned personally, she "understood" both the desperately disfiguring reality of sin and the wonderfully redemptive blessing of the forgiveness of sins. Furthermore, Mary knew, too, the sublime beauty of innocence. Mary knew not only the "original" innocence that God intended for each one of us, but she also knew it as a *redemptive gift* of her Son, Jesus Christ; and, on that basis, she also understood how wonderful was the work which God was accomplishing in her (cf. Lk. 1: 49) and, indeed, did accomplish in her.

Finally, then, our hope is in Mary on both counts. On the one hand she knows how to cooperate with grace. Just as she cooperated with the grace of the *Immaculate Conception*, so are we called to take up the gift and task of the sacrament of Baptism. Therefore, Mary is able to understand, to encourage and to help us with our graced cooperation with the "redemptive gifts" of her Son. On the other hand, as she understood sin from the point of view of the harm it did to her Son, she understands more fully the tragedy that it is and the depths of forgiveness which await the sinner. Moreover, just as a mother's love is ever hoping in the return of the prodigal son, so is her prayer effective for our final perseverance.

Conclusion.

In the first place, Mary is a sign of hope in that God *accomplished in, with and through her*, what he seeks to accomplish *in, with and through us*.

Secondly, we see in Mary that the full meaning of to "hope against hope" is *to hope in Christ for our salvation*. For 'She stands out among the poor and humble of the Lord, who confidently hope for and receive salvation from him'[44]: a salvation that is impossible for us to obtain for ourselves. Thus "hoping against hope" for salvation is pre-eminently knowing that it cannot come from the "self"; rather, salvation can only come from the paschal grace of Jesus Christ.

Thirdly, it is the plan of God that Christ gives us His mother to be our mother; and what good (cf. Gn. 1: 31) mother does not *actively seek the good of her children?* Thus 'Mary can be said to continue to say to each individual the words which she spoke at Cana in Galilee: "Do whatever he tells you"'[45]. For Christ is our good. And thus we can *hope in the ministry of God's choice of her maternal love of us.*

Fourthly, Mary knew from her faith[46] experience the work of the Blessed Trinity in her life[47]; and, therefore, she is in a unique position to educate[48] us in our relationship to the Blessed Trinity. For Mary is the "daughter of the Father"[49], the "Mother of the Son" and the "spouse of the Holy Spirit"[50]. In

---

[44] *Lumen Gentium*, 55.

[45] *Redemptoris Mater*, 46, p. 144.

[46] CCC, 2005.

[47] Cf. Hans Urs von Balthasar, *Mary for Today*, translated by Robert Nowell, Slough: St. Paul Publications, reprinted 1989, p. 35.

[48] St. John Paul II, (Karol Wojtyla), *The Way to Christ: Spiritual Exercises*, translated by Leslie Wearne, Harper San Francisco, 1994: Chapter 4: A Talk for Female Students, p. 37: 'Her basic task is that of educating, and when she shares the responsibility for it with men she cannot be simply an object for them.'

[49] *Lumen Gentium*, 53.

[50] Cf. *A New Marian Dogma?*, by Thomas Xavier, (a review of *Mary: Coredemptrix, Mediatrix, Advocate.*, edited by Mark I. Miravalle), on p. 62 of the May 1997 issue of the magazine *Inside the Vatican*, edited by Robert Moynihan. Cf. also CCC, 505-507.

other words, if ever there was a *unique* relationship of a creature to their Creator-God, then it is that of the Blessed Virgin Mary to each and to all the persons of the Blessed Trinity.

Finally, there is a very particular sense in which *our hope is in Mary*; and this is because if God *chose* to come to us *through the Virgin Mary*, then He chose us to come to Him through the same Virgin Mary. Should we not seek God through the very same 'Gate of Heaven'[51] that God Himself chose in order to seek us?[52] In other words, the "woman" is an important, even an indispensable "word" for man. Therefore, somewhat astonishingly, salvation is as much about God restoring the right relationship between man, male and female, as it is about reconciling us to the Father. In view of the actual nature of the events of salvation history, perhaps that right relationship to God *begins in the right relationship* of man to woman, which begins with gratitude. On the one hand, there is a man's gratitude to woman for the vocation of motherhood (cf. Tobit, 4: 3-4); and, on the other hand, there is a man's gratitude to Mary for her full participation in being the 'mother of all [the] living' (Gn. 3: 20). Thus, notwithstanding the mystery of the original creation of man, male and female, the dependence of "man" on woman is a sign of salvation from God; and, in consequence, seeking to be freed from that dependence is a sign of pride: of "man's" desire to be self-sufficient.

In the end, then, the actual nature of our salvation is what *befits* the real nature of our condition; and, in view of the immense blessings of salvation, blessed be God for the blessings of the Mother of Our Lord Jesus Christ!

---

[51] *A Simple Prayer Book*, p. 15.

[52] This is not a contradiction of the fact that our redemption is from Christ, for this holds of Mary, too, as has been stated in the course of this article; rather, this is an exposition, as it were, of the fact that God chose to become a man born of a woman and, in so doing, took up the "maternity" of this woman not just for Himself but for all of us.

FOREWORD TO CHAPTER FOUR

EDMUND ADAMUS[1]

As I read through this remarkable chapter on 'Mary and Prayer' the following sentence began to linger in my mind. "The nature of the suffering between Christ and His Mother exists as a kind of relationship within which it is possible for all human experience to find a home." We know that from deep within our Catholic tradition, the devotion to both the Sacred Heart of Jesus and the Immaculate Heart of Mary rests upon the ability of the Christian soul to see in those two Hearts a place they can go to for comfort. A place to take spiritual 'rest' from the turmoil and tribulation of life's suffering and chaos. It is precisely because Jesus never ceased in his prayer to the Father right up to his last breath, and because Mary 'pondered all these things in her heart,' that the disciple can bring his or her own prayers to the first and true disciple of Jesus, Mary who in turn re-presents our petitions to the Merciful Heart of her son. St. Maximilian Kolbe once wrote that "the Immaculate Mother of God is the ladder upon which we all climb to reach the Sacred Heart of Jesus."

Mary: "step by step" to her Son

I rather like this image of Mary as a ladder. After all, what is a ladder but a useful apparatus by which we can reach something higher, it's also an

---

[1] *Edmund is Education Consultant for www.fertileheart.org.uk a new moral formation curriculum for Catholic schools . !*

effective instrument by which we can see something more clearly, it can give us a new and better standpoint from which to take in a different and more accurate perspective of what we need to do. Think of how much easier it is to paint a ceiling with a ladder or repair something high up and lofty when we have the right kind of ladder.

We also know that a ladder is only as good as the rungs upon which we place our weight to reach whatever it is we have to get to. A ladder can only fulfil its purpose with safety and security if the bottom of it is placed firmly on the ground so that when we ascend step by step we can trust it to remain steadfast and reliable. Therefore, when one thinks of praying with and through Mary to draw closer to Christ, one can see how the image of her as a 'ladder' is a rather useful one. We too need a firm, steadfast faith to begin our ascent in prayer to Jesus. The 'rungs' of our climb are the things we might call to mind about the life and faith of Mary and Jesus. In this sense the great mysteries of the Holy Rosary might well be viewed as the 'rungs' or steps of the ladder [Mary] which in contemplating them one by one they immerse us more deeply in to the search for the Sacred Heart of Jesus, the Heart whom Mary knows more intimately than anyone who has ever lived. Indeed, in the recitation of the rosary even the individual beads within each decade might well be the tiny but significant "steps" upwards and upwards from the base of our doubt, despair and lack of faith slowly and surely towards the light and hope of grace at the top of the climb that comes through patient perseverance in the praying and recitation.

Mary's "maternal arm around us"

So when we invoke the intercession of Mary we are as it were placing our weight, and these burdens maybe exceedingly great and heavy from an emotional and spiritual perspective, upon the Heart of our Mother who will not let us down. She above all knows what suffering and pain and desolation feels like. She wants to help carry our sadness and woes alongside us because just like the transformation of her acute human pains in life and indescribable grief at the suffering and death of her beautiful son in to sorrow, she can help turn our grief in to hope-filled sorrow. It won't make the source of the pain or anxiety evaporate but by the strength and power of her maternal arm around us, the pain can lessen, and at least become more bearable as we seek out that "home" of comfort and consolation in the bosom of Christ, just as John the Beloved Disciple laid his head to rest upon the Lord at the Last Supper.

"I experienced a profound and real sense of God's love and mercy through Mary"

I have been privileged to have known this special comfort at various times in my life but none more so than on the occasion of a particularly significant ecclesial event. During 2012 I was part of a unique visit of a replica image of the icon of Our Lady of Czestochowa to Great Britain known as "From Ocean to Ocean." www.fromoceantoocean.org It began as an act of entrustment of the civilisation of love and life in to the hands of the Mother of God and it was given the title 'From Ocean to Ocean because it was a pilgrimage of the icon of the Mother of God from the Pacific Ocean at Vladivostok to Fatima in Portugal and the Atlantic. This was a pilgrimage through more than 30 nations of over 18,000km. I was part of the team that hosted the icon from 5th-16th November, arranging it to be received in Kent to Westminster, to Walsingham, Birmingham, the North West, then eventually to Scotland and over the sea to Ireland. It was during the celebration of Holy Mass at the shrine of Our Lady of Walsingham when the icon was being venerated that I experienced a profound and real sense of God's love and mercy through Mary.

For years I had been struggling to really forgive my father and mother, both of whom had died years before for injustices I felt I had suffered at their hands. I loved them and prayed for them but there was still a blockage in me that was holding me back. My wife and I lost our first son, Patrick in 2008 at 7 months stillborn. It was a devastating cross. That day during the Mass and before the holy icon, I prayed with my eyes closed listening to the singing and praise and I experienced an inner vision from deep within me. Slowly but surely I could see two figures approaching me. Eventually I could see that they were my parents, looking elderly but not frail with old age and sickness as I had last seen them. They were glowing with health and joy and then as they drew closer, I could see that they were holding my child Patrick in their arms; embracing him and kissing him tenderly. I knew from that instant that my parents were in eternal peace and my son was with them. I knew from that moment that I could and had forgiven my parents from my heart. And I knew that it was the Immaculate Heart of the Mother of Christ who had granted me this special grace as some reward for the many hours and days of hard and demanding work I had devoted to the success of the British leg of the Ocean to Ocean pilgrimage.

So yes, Mary is as St. John Paul II reminds us a "school" and that day she lovingly bestowed on me a lesson I shall never forget and for which I will always be grateful for the peace it brought to my soul.

Chapter Four

Part III of a Marian Triptych: Mary and Prayer

General Introduction to Chapter Four: Prayer. 'The Fathers of the Church say that prayer, properly understood, is nothing other than becoming a longing for God. In Mary this petition has been granted: she is, as it were, the open vessel of longing, in which life becomes prayer and prayer becomes life'[1]. 'We can ask ourselves: when I pray, do I open myself to the cry of so many close and distant persons? Or do I think of prayer as a sort of anaesthesia, to be able to be more tranquil?'[2]

Thinking of prayer, then, it can seem that it is about "moments of prayer" as if, owing to the crises of life, we turn to God in prayer because it is impossible that anyone else can understand or help us; and, indeed, this remains a consolation in the depths of human suffering: that the aloneness we experience is a kind of *aloneness with the alone*[3]. I particularly remember an intensely difficult time in my married life when the meeting in which this

---

[1] Joseph Ratzinger, p. 15 of *Mary: The Church at the Source*, 2005.

[2] Pope Francis, "Pope at General Audience on Jesus' way to Pray (Full Text)", Virginia Forrester, February 13th: https://zenit.org/articles/pope-at-general-audience-on-jesus-way-to-pray-full-text/.

[3] Cf. A song composed by Kiko Arguello of the Neocatechumenal Way is called 'Sola a Solo' (Amsterdam, 30th April, 2005): 'Sola a Solo, under the cross, Mary, who can separate you? Virgin alone …'. Cf. also Francis Etheredge, *The Prayerful Kiss*, from enroutebooksandmedia, 2019: http://enroutebooksandmedia.com/theprayerfulkiss/.

song was sung and shared gave me an opportunity to talk about the pain that existed within; and, almost on the basis of solely being listened to, the experience of this pain being heard seemed to lance and help it. There are depths to the human heart, then, that we almost cannot reach in the human terms of analysing and explaining; but, in the experience of "being with", there is a kind of transcendence of the isolating, almost socially suffocating sense of being unheard by others.

In the intense experiences that people go through in the prolonged experience of pain, whether it is the pain of infertility, of the suffering inflicted or otherwise brought about by others, the length and depth of an illness, there is a kind of point of darkness that is almost out of reach and incapable of being communicated; and yet, this interior pain is not unintelligible and belongs to the profoundly shared human experiences which are, in a way, the current that runs through human nature and throughout the whole "liturgically real" experience of the paschal death and resurrection of Jesus Christ. In other words, although it seems to us at different times that there are experiences which are "unshareable", the nature of the suffering between Christ and His Mother exists as a kind of relationship within which it is possible for all human experience to find a home. In the difficult if not impossible experiences which often constitute the human kernel of a "bioethical" crisis, dilemma or apparent dead end, there is a prayer which rises out of the depths of the human spirit which is touched by the Spirit of God: a prayer which opens and engages in a wordless dialogue with God in a way that makes help possible.

Chapter Four: Introduction. Mary is not a conceptual counterpoint to Christ, imagined and projected into the Gospels as a kind of semi-divine goddess who was made manifest, as it were, in preparation for the coming of the Saviour of mankind: a kind of "power of woman" to obtain what she wants. Even if, then, in the way of human frailty, Mary bears the imagined possibilities of a woman out of time, before her time, independent of time, it is necessary to consider what has made possible the unprecedented growth of Christianity's relationship to Mary, Mother of the Lord: 'it is admitted that the praises of Mary grow with the growth of the Christian community, we may conclude in brief that the veneration of and devotion to Mary began even in the time of the Apostles'[4]. In addition, given the general sense of inequality between men and women, evidenced in men marrying more than one woman[5], men's recourse to divorce (Mt. 19: 3, 7-9) and the woman's burden of blame for sin (cf. Gn. 3: 12; Jn 8: 3-5), it follows that there is a work of God in the slow, steady and stable disclosure of the greatness of His gift of Mary in the history of salvation and the life of mankind. Given, furthermore, the human tendency to go to a person who will positively influence our request, just as a child goes to his mother, it follows that if grace builds on nature then it is "supernaturally natural" to turn to the maternal help of the Mother of God, Mother of the Church and our Mother.

Thoughts on Mary and prayer, then, could easily begin with specific Marian prayers, like the Rosary, but the problem here is that Marian prayer will look like an additional type of prayer to the prayer of the Church; but if, more globally, the point of departure is the relationship between Mary and the Church, then it follows that Mary and Prayer are as intimately ordered to one another as breathing is to the body. In the third and final piece of this triptych on Mary it is necessary to consider the following: The Emergence of Mary (I); Mary and the Prayer of the Church (II); and Marian Prayer and the Covenant (III).

---

[4] Anthony Maas, "The Blessed Virgin Mary", *The Catholic Encyclopedia*, vol. 15, New York: Robert Appleton Company, 1912, 10 Jan, 2019: http://www.newadvent.org/cathen/15464b.htm.

[5] "Polygamy": https://www.catholicculture.org/culture/library/dictionary/index.cfm?id=3562 1.

The Emergence of Mary (I)

'Throughout the Old Covenant the mission of many holy women prepared for that of Mary'[6]. Mary is referred to in the opening of St. Matthew's Gospel with the brief words: 'and Jacob the father of Joseph the husband of Mary, of whom Jesus was born, who is called Christ' (Mt. 1: 16). At the same time, however, there is no doubt that the identity of Christ required an explicit and detailed witness to bring out the mysterious nature of the *Incarnation*, the Eternal Word taking on the wordless[7], embryonic flesh, from the flesh of Mary: 'And the Word became flesh and dwelt among us, full of grace and truth' (Jn. 1: 14); and, therefore, the infancy narratives in St. Luke's Gospel, complemented by the early parts of St. Matthew's Gospel, are almost wholly directed to communicating the unique nature of Christ's conception and the revelation of Mary's exceptional expression of God's preparation for her active acceptance of being 'the handmaid of the Lord' (Lk. 1: 38).

Early Christian Tradition, however, has both recognized Mary as being prophetically anticipated (cf. Gn. 3: 15, 20) and conceived through the spousal love of Anna and Joachim; and, indeed, it is held that Mary was the fruit of their 'fervent prayers' for a child[8]. Elizabeth, through the impulse of the Holy Spirit, at once identifies Mary as the 'mother of ... [the] Lord' (Lk. 1: 43) to which Mary exclaims emphatically: 'My soul magnifies the Lord, and my spirit rejoices in God my savior' (Lk. 1: 46-47). In other words, Mary

---

[6] *The Catechism of the Catholic Church*, (CCC) 489: Eve; Sarah; Hannah; Deborah; Ruth; Judith and Esther; and many other women; and cf. also, Cardinal Ratzinger (now Pope Emeritus Benedict XVI), *Daughter Zion*, translated by J. M. McDermott, San Francisco: Ignatius Press, 1983.

[7] From a Homily by Fr. Anthony Trafford, 2019: 'infant' translates as 'without words'; cf. also 'infant': https://www.etymonline.com/word/infant.

[8] Cf. Anthony Maas, "The Blessed Virgin Mary", *The Catholic Encyclopedia*, vol. 15.

prophetically announces and fulfils her vocation in giving witness to God 'who is mighty [and] has done great things for me' (Lk. 1: 49); and, in that sense, identifies the vocation of the Church as recognizing, proclaiming and praising the works of God.

Mary, it is clear, is coming into increasing prominence not least of which is due to the doctrinal clarifications that she was conceived without original sin, assumed body and soul into heaven and that she expresses, in a unique way, the "personal subject" of salvation who is "re-conceived" in Christ and, therefore, re-conceived in the relationship between Christ and His Church. Mary 'is hailed as pre-eminent and as a wholly unique member of the Church, and as its type and outstanding model in faith and charity' (*Lumen Gentium*, 53). Just, then, as Mary is 'pondering ... [all these things] in her heart' (Lk. 2: 19), so does the Church and, therefore, each one of us participate in the vocation to ponder the works and words of God (cf. *Dei Verbum*, 2). Thus there is a kind of natural origin to the prayer of the Gospels emerging, as it were, out of pondering a variety of moments in the life of Christ: 'She withstood [the poverty of the manger ... the loss of Jesus as a boy in Jerusalem ... and finally the death on the Cross of her beloved Son] ... through faith and hope, meditating in her heart the joyful and painful mysteries of the history of salvation'[9]. In other words, what is expressed "exteriorly" as the Marian prayer of the rosary is a kind of outward expression of the "interiority" of the events which Mary pondered in the life of her Son, Jesus Christ[10]. In the development of this prayer, perhaps we can speak of a Marian inspiration in the Church taking further the "moments" that illuminate the Christian life in the Mysteries of Light (cf. *Rosarium Virginis Mariae*, 3); indeed, even though 'In these mysteries[11], apart from the

---

[9] Fr. John, Saint Mary of Lagrasse Abbey (France), "Throughout her life, Mary's love was put to the test" (A Moment with Mary [lettre@mariedenazareth.org]).

[10] Cf. St. John Paul II, *Rosarium Virginis Mariae*, 11.

[11] '(1) his Baptism in the Jordan, (2) his self-manifestation at the wedding of Cana, (3) his proclamation of the Kingdom of God, with his call to conversion, (4)

miracle at Cana, *the presence of Mary remains in the background*' (*Rosarium Virginis Mariae*, 21), perhaps we can see an image of the "humility of Mary and the Church" in her proclamation of the Christian mysteries.

Mary and the Prayer of the Church (II)

> 'There are some who think that the centrality of the Liturgy, rightly stressed by the Second Vatican Ecumenical Council, necessarily entails giving lesser importance to the Rosary. Yet, as Pope Paul VI made clear, not only does this prayer not conflict with the Liturgy, *it sustains it*, since it serves as an excellent introduction and a faithful echo of the Liturgy, enabling people to participate fully and interiorly in it and to reap its fruits in their daily lives' (St. John Paul II, *Rosarium Virginis Mariae*, 4).

But if Mary is the 'type' of the Church when the Church prays it is praying the prayer of Mary; and, therefore, Marian prayer "brings out", as it were, the nature of human prayer prayed in the Holy Spirit. Hans Urs von Balthasar says that the *'"Hail Mary!" is a training and an integration into the prayer of the Mary-the-Church'*[12] and then goes on to say that *'The official liturgical prayer of the Church, too – whether openly or hiddenly, consciously or unconsciously – is always Marian Prayer'*[13].

On the one hand, the existence and action of the Holy Spirit in the nature and life of the Church is 'compared by the Fathers to the function that the principle of life, the soul, fulfils in the human body' (*Lumen Gentium*, 7). Thus Mary, who is the 'type' of the Church is, as it were, praying the prayer

---

his Transfiguration, and finally, (5) his institution of the Eucharist, as the sacramental expression of the Paschal Mystery' (*Rosarium Virginis Mariae*, 21).

[12] In *The Threefold Garland*: The World's Salvation in Mary's Prayer, translated by Erasmo Leiva-Merikakis, San Francisco: Ignatius Press, 1982, pp. 22-23 (italics in the original text).

[13] Balthasar, *The Threefold Garland*, p. 23 (italics in the original text).

of the Holy Spirit; indeed, the Church herself says of Mary that she is 'the perfect *Orans* (pray-er), a figure of the Church' (CCC, 2679).

On the other hand the woman, Mary, is in the midst of human relationships in the way that the Church is both immersed in human reality and distinct from it: the Church is both 'the visible society and the spiritual community' .... [and forms] 'one complex reality which comes together from a human and divine element' (*Lumen Gentium*, 8). Thus Mary, in all the ways that the Church is immersed in the human, social and ecological reality of salvation history receives, takes up and prays the prayer of the Christian Church in her dialogue with Christ: 'When the wine failed [at the marriage feast of Cana], the mother of Jesus said to him, "They have no wine"' (Jn. 2: 3). Thus Mary first turns to Christ and then to us, saying, "Do whatever he tells you" (Jn. 2: 5). Thus St. John Paul II comments: 'The revelation made directly by the Father at the Baptism in the Jordan and echoed by John the Baptist is placed upon Mary's lips at Cana, and it becomes the great maternal counsel which Mary addresses to the Church of every age: "Do whatever he tells you" (*Jn* 2:5)' (*Rosarium Virginis Mariae*, 21).

Marian Prayer and the Covenant: The Conversation of Conversion (IIIi)

When pondering, then, the relationship of Mary and prayer, it is not so much that there is Marian prayer; rather, it is as if all prayer is Marian in that all prayer expresses the relationship of creature to Creator. Thus the following question arises. If Christ is true man and true God - Is the prayer of Christ "Marian prayer"? If Christ is true God then His prayer is the prayer of the Holy Spirit; but, if Christ is true man, then His prayer is the prayer of the Holy Spirit. But there is, too, an ontological dialogue, as it were, which founds the relationship between the humanity of Our Lord and Savior Jesus Christ and His Blessed Mother Mary: 'Womanhood and manhood are complementary *not only from the physical and psychological points of view,* but also from the *ontological*[14]. Thus the question arises: What is the ontological complementarity that is both naturally founded and, therefore, a work of God - the spiritual fruit of which is expressed in the dynamic relationship between Christ and His Church? On the one hand, there is the eternal Word of God who is "made flesh" – making "flesh" the word of God through a kind of "incarnation" in the life of Christian men and women; and, on the other hand, Mary is the woman of faith: the one in whom the Word of God took flesh and the word of God found a fruitful home. Mary, therefore, '"precedes" everyone on the path to holiness; in her person "the Church has already reached that perfection whereby she exists without spot or wrinkle (cf. *Eph* 5:27)". In this sense, one can say that the Church is *both* "Marian" and "Apostolic-Petrine"'[15]. In other words, from a variety of intersecting states, vocations and gifts of God, Mary transpires to be in the heart of the redeemed creation; and, therefore, in the reality of human relationships, it is possible that being a "man" entails learning to appreciate "woman" in the 'school of Mary'[16].

Prayer, in that it is a relationship between the *Persons* of the Blessed Trinity, is an expression of the *relational* nature of God; indeed, in the words of *The Catechism of the Catholic Church*: 'The heart .... is the place of encounter, because as image of God we live in relation: it is the place of covenant' (CCC, 2563). Thus Mary, placed as she is by God Himself in the midst of the Blessed Trinity, is in the midst of this divine and human dialogue: 'The angel announced to ... [Mary] not just the incarnation but

---

[14] Pope St. John Paul II, *Letter to Women*, 7.

[15] St. John Paul II, *Mulieris Dignitatem*, 27.

[16] Cf. St. John Paul II uses this expression, 'the school of Mary', in various documents; and cf. more generally the following article for the idea of a relationship to Mary helping men: Keith Berube, "Mary Is Made for You", https://www.hprweb.com/2018/12/mary-is-made-for-you/ .

fundamentally the entire mystery of the Trinity'[17]; and, if Mary is in the midst of this dialogue, then we too are taken into the heart of this encounter: this place of meeting in which all human loves and concerns are brought into the conversation of conversion with the living God. In obedience to God, then, 'Mary places herself between her Son and mankind in the reality of their wants, needs and sufferings'; and, at the same time in doing so, she is pre-eminently placed to wish 'the messianic power of her Son to be manifested, that salvific power of his which is meant to help man in his misfortunes, to free him from the evil which in various forms and degrees weighs heavily upon his life'[18].

The Covenant and Prayer (IIIii)

'Baptism is prefigured in the crossing of the Jordan River by which the people of God received the gift of the land promised to Abraham's descendants, an image of eternal life' (CCC, 1222); and, furthermore, 'The promise of this blessed inheritance is fulfilled in the New Covenant' (CCC, 1222). On the one hand, 'Prayer is lived in the first place beginning with the realities of *creation*' (CCC, 2569); and, on the other hand, 'In his indefectible covenant with every living creature, God has always called people to prayer' (CCC, 2569).

The 'New Covenant', then, is instituted by Christ and, as it were, together with the disciples, constitutes the Church (cf. Lk. 22: 20) in a "Eucharistic attitude": 'When Mary exclaims: "My soul magnifies the Lord and my spirit rejoices in God my Saviour", she already bears Jesus in her womb. She praises God "through" Jesus, but she also praises him "in" Jesus and "with" Jesus' (*Ecclesia de Eucharistia*, 58). Thus Christ fulfils the interpersonal gift that

---

[17] Hans Urs von Balthasar, *Mary for Today*, translated by Robert Nowell, Slough: St. Paul Publications, p. 35.

[18] St. John Paul II, *Redemptoris Mater*, 21.

God gives, from the beginning, to make it possible for us to exist in a saving relationship to Him; and, therefore, from creation to the first rainbow covenant with Noah and 'the earth' (Gn. 9: 13), through the covenant of the circumcision of Abraham and his people (cf. Gn. 17), to the covenant of the heart (cf. Heb. 10: 16) and the coming of Christ: the covenant expresses and enacts the saving relationship of God with His people. Baptism, then, enters us into the New Covenant: 'From the baptismal font is born the one People of God of the New Covenant, which transcends all the natural or human limits of nations, cultures, races, and sexes: "For by one Spirit we were all baptized into one body" [1 Cor. 12: 13]' (CCC, 1267). Thus: 'In the New Covenant, prayer is the living relationship of the children of God with their Father who is good beyond measure, with his Son Jesus Christ and with the Holy Spirit' (CCC, 2565).

The Experience of Prayer "Beads" (IIIiii)

In what follows there will be a brief selection of experiences with reference to prayer, spanning not much short of sixty years and many different periods and phases of life. These moments are not so much about how "efficacious" they were as that they reflect specific times that "prayer" almost sprang into existence, then subsided, before returning to the Church re-established the regular habit of prayer; and, as prayer has become like the Christian breathing it is, prayer has not only developed and become, incredibly, almost constant – but that it also takes up more and more of the nature of daily life until life and prayer seem to run into and from one another.

As an adolescent on a farm, then, driving a tractor in the fields and out of earshot, I can remember singing the praise of God with all the energy of youth; and, therefore, even without knowing it there arose in me the 'Eucharistic attitude': an attitude that seemed to spring from nowhere. On reflection, however, maybe this 'Eucharistic attitude' sprang from the roots of baptism, however overgrown or diseased it was; and, even if many years were to pass before I remember praying again, there remains a delight in the beauty of nature which is often the start from which a spirit of praise takes a beginning.

I was once lost, being in my twenties and having just visited a monastery in Northern Ireland in which I experienced a kind of intimation that my life would involve service at "the kitchen sink"; and, although I do not remember distinctly, there was a sense of "being told" the name of a pub to ask for as I was hitchhiking away without really knowing where I was going. When I arrived at this pub there were very few people there and the barman invited me to wait and see what would happen. As the evening wore on a group came in, one of whom I recognized as someone I had waited on in a Hotel where I had once worked, briefly, and he and his wife gave me accommodation for the night. Even if, however, I was rebellious or incredulous at the time it is nevertheless true that family life is deeply rooted in the vocation to practical service and, although many years later, it is clear that prayer, service and humility are inseparable in both marriage and parenthood. While driving all over the country as a portacabin repair man I remember praying and, as the horizon came into view, almost hearing the words of the Gospel: 'what does it profit a man, to gain the whole world and forfeit his life?' (Mk. 8: 36). In my late twenties, then, I abandoned both the life of a journey man and set off for spiritual direction. One outcome of these visits to a monk in London was the possibility of entering a religious order in Scotland; however, I rebelled against this and hitchhiked back to the Abbey from which I had set off. Although I set off in the early evening, from a relatively isolated place with little prospect of a quick journey, the time it took for me to travel from a remote place in Scotland to the monastery in London was "as long as it took me to say a rosary". None of this was immediately fruitful in terms of "coming to my senses" (cf. Lk. 15: 17) and knowing what my life was about; but, nevertheless, these moments were distinct "beads" in my relationship to prayer.

Prayer, Conversion and the Help of the Holy Family

Pope Francis said: "Conversion ... is the first step of healing in the sense that it opens the heart so that the Word of God may enter"[19].

Prayer as I prayed it, however, was intermittent and rooted in the situations of life which, as real as they were, were a part of what was, really, a faithless Christian life; and, at forty, exhausting many attempts to marry, the paining loss of a child to abortion, the ongoing inability to find a vocation, to recognize and fulfil my human talents, I was again turning over the possibility of suicide in front of the reality of being a sinner. Into this abysmal moment came the words of the *Catechism of the Catholic Church* and, immeasurably more personally significant for me, I believed them: 'Since God could create everything out of nothing, he can also, through the Holy Spirit, give spiritual life to sinners by creating a pure heart in them' (CCC, 298). Turning to God in a new way brought me to recognize my vocation to marry and to be open to life. Ten children later, two of whom are in heaven with a third from earlier on in my history, the whole gift of being open to life is an expression of the gift of believing in the God who helps; and, at the same time, just as there are major crises so there are minor crises too, including the practical needs of everyday life. I remember trying to hang a towel rail for any number of towels and finding that no matter what I did to the rail it fell off. A prayer to St. Joseph, however, helped me to find an answer in the invention of a towel rail. I used two flower pot hanging basket brackets, one at either end of where we needed the towels; and, slotting light metal piping into the curls of the flower pot brackets, lo and behold we had a rail that would take several towels. What is especially helpful is that the whole towel

---

[19] Deborah Lubov, 7/2/2019, "Pope's Morning Homily: Conversion Heals, Creates Us Anew":

https://zenit.org/articles/popes-morning-homily-conversion-heals-creates-us-anew/.

rail system has stayed on the wall!

In the beginning of married life, rooted as it was in the vocation to draw on the whole Christian life, one of the first mysteries of the Rosary to attract my heart was the marriage feast of Cana in which Christ turns water into wine: our sufferings into joys. In the words of Pope Francis: 'The Scriptures, especially the Prophets, indicated wine as a typical element of the messianic banquet (Cf. *Amos* 9:13-14; *Joel* 2:24; *Isaiah* 25:6). Water is necessary to live, but wine expresses the abundance of the banquet and the joy of the celebration. A celebration without wine? I don't know . . .'[20]. There are different kinds of wine: the answer to a prayer when we are in financial difficulties, the finding of another washing machine when our old one burnt out and coming across people who will collaborate with writing projects; the sudden and unexpected breakthrough in the difficulties we have in transmitting the Christian Faith to our children, with their friendships and with helping them with their interests and opportunities; and, finally, the joy of being forgiven, of a word of God which helps us and the gift of time together in our Church-Communities. As the prayer continues to be prayed over the years and in every situation of marriage and family life so each of the Christian mysteries began to attract, as it were, intentions and experiences that continue to grow with it.

Thus there is a growing realization that the prayer that arises out of our lives, real as it is, needs to be rooted in the reality of conversion and the turning to God that this entails; and, by implication, conversion enters more and more centrally into the relationship of God to man becoming, almost, the central axis of life through which all else turns. This is not to diminish the reality of our "moments" of prayer; but, rather, looking at prayer in the light of the history of salvation, it is an essential part of this history that we

---

[20] "Angelus Address: On the Miracle at the Wedding of Cana, 'Do Whatever He Tells You'", January 20th, 2019, translated by Virginia M. Forrester, https://zenit.org/articles/angelus-address-on-the-miracle-at-the-wedding-of-cana-do-whatever-he-tells-you/.

pray: because prayer takes us into the relationship with 'the living God who wants men to live' (CCC, 2575). The prayer that develops, then, with all its family facets, crises, the needs of others, social concerns and practical needs, is more like a growing presence of the Holy Family and their involvement in our daily life[21]. Just as prayer draws on the history of salvation so it invites us, as it were, to share in the process of pondering that history; and, at the same time, these iconic prayer beads become, as we pray, like living illuminations of the kaleidoscopic moments in both the prayer of the Holy Family and our own family life.

The Power of One Hail Mary

I remember one simple "Hail Mary" while all ten of us were on pilgrimage to *The World Meeting of Families* in Milan (in 2012). We had left the main group and gone for a walk and, after one of our children accidentally flung one of her trainers into a lake and we were unable to retrieve it, we discovered the main group had gone and, after going in search of a bus back to our host accommodation, that we were also lost. Standing in a bus-shelter, aware of an ants' nest and the pushing and shoving of tired children, we prayed one "Hail Mary". At the end of this prayer an Italian drove up who recognized us from the meeting earlier that day but who spoke no English; he got out of his car and spoke to the driver of a bus that had stopped, who then let all ten of us on the bus, free of charge and took us to the Bus-Station. The Italian followed the bus and made sure that when we arrived at the Bus-Station that we got on the right bus back to our accommodation! Guardian angels obviously do not need to speak English!

---

[21] Cf. Francis Etheredge, *The Family on Pilgrimage: God Leads Through Dead Ends*, St. Louis: En Route Books and Media, LLC, p. 1: http://enroutebooksandmedia.com/familyonpilgrimage/.

FOREWORD TO CHAPTER FIVE

Dr. Michal Pruski

Procreation – a miracle in the mundane

Bringing forth a new life seems to be simultaneously one of the most amazing and mundane acts a human can participate in. It is mundane in many ways. It is not so much an act of our intellect but of the animal part of our nature, where we respond to one of our most basic inclinations,[1] an act that the brutes perform as well. Humans were performing it since we walked the earth, and scientists have studied it and described it in the driest way that scientific accuracy, in textbooks and the peer-reviewed literature, allows for. On the other hand, it is an amazing act in which two creatures partake in God's creative power. When a husband and wife join in the marital act, with the blessing of God, they do not just create a thing, but are together with God co-creators of a new living organism, a person – soul and body – recognising the good their act will bring about. This new human being will see in their conception the beginning of their existence – for we began as persons at conception and not at birth.

Mary's conception was even more of a wondrous event, for here in the fallen world God allowed someone to be begotten without the stain of Original Sin – a fact which the Church proclaimed as a dogma.

---

[1] *Summa Theologiae*, hereafter: STh, I-II q94 a2.

Ensoulment – still a mystery

In this chapter Etheredge explores this dogma of the *Immaculate Conception* and how it can illuminate our understanding of our own beginning. As this chapter states, the Church has proclaimed very little with regards to the beginning of human life. The Church has never stated when ensoulment happens,[2] though such an authoritative opinion as that of St Thomas Aquinas suggests that it happens during conception.[3] Moreover, conception is itself a much more complex phenomenon than many would think it to be. Etheredge notes that when talking about conception we could either talk about the beginning of a new life when the sperm penetrates the egg or when the two pronuclei of these gametes fuse together – a discussion known to Bioethics, but not one that has received a lot of attention.[4] While we know that the two gametes in themselves are not individual persons,[5] and that the status of the embryo, with its unified genome is certainly that of a distinct individual human being, the status of the just

---

[2] Cf. Jones DA. *The Soul of the Embryo: An Enquiry into the Status of the Human Embryo in the Christian Tradition*. London; New York: Continuum; 2004 & *Evangelium Vitae*, para 61.

[3] Aquinas' account of ensoulment is rather complex and not without a critique, especially with regards to the three stages of it (STh I q118 a2), though in STh III q33, where he discusses Christ's conception, he states that it is animated instantly with a rational soul. For a discussion on this issue see Eberl JT. *Thomistic Principles and Bioethics*. 1 edition (Kindle). Routledge; 2006, particularly chapter 2 and footnote 12 to chapter 3 & Jones DA. "The human embryo in the Christian tradition: a reconsideration". *Journal of Medical Ethics*, 2005; 31:710–4.

[4] Leone S. "The Pre-zygote Identity as a Moral Issue". *Human Reproduction & Genetic Ethics*, 2008; 14:15–21.

[5] Cf. Mackellar C. "Representative Aspects of Some Synthetic Gametes". *The New Bioethics*, 2015; 21:105–16 & Pruski M. "The Relationship of Gametes to Those Who Procreate and Its Impact on Artificially Generated Gamete Technologies". *Ethics & Medicine*, 2017; 33:27–41.

penetrated ovum is not clear. To help to shed light on this issue Etheredge utilises the fact that we owe *Obsequium religiosum*[6] – religious submission of the will and mind - to the Church's pronouncements on the matters of faith and morals. So as much as we know that reason can enlighten our faith,[7] with regards to ensoulment and the beginning of human life our faith can enlighten our reason. Through considering what we know about Mary's beginning we can learn something about our own beginning. By looking at what occurs when the sperm penetrates the ovum and what the Church teaches regarding Mary's Immaculate Conception, Etheredge provides a compelling case for why we should regard this moment as the moment of ensoulment. Of course, the matter has not yet been settled, and counterarguments are possible. Yet, with the speed of scientific progress, the Church cannot afford to leave the issue unsettled.

Guiding questions to the debate

One might perhaps wonder what is the purpose of this debate and why is this an urgent matter for the Church to resolve? Surely the time difference between the union of the sperm and the egg, and the union of their pronuclei is very small.[8] So could this time difference be of any significance? As Catholics, we know that conception should happen in the marital act and not in the *in vitro* environment; we know that artificial methods of contraception and abortion are wrong - we should neither hinder the coming to be of a new life, nor the development of an already existing human being. Will answering this question do anything more than

---

[6] *Lumen Gentium*, para 25.

[7] Cf. *Fides et Ratio*.

[8] Leone S. "The Pre-zygote Identity as a Moral Issue". *Human Reproduction & Genetic Ethics*, 2008; 14:15–21.

indulge an obscure academic debate?

The importance of a definition

There is a deep need in us to trace our beginning. This not only materialises itself through the desire of adopted children to know their biological parents, a desire to know the history of our families, but also to answer an innate existential question: when did I begin?[9] To answer this question we need to know what signifies this moment.

Moreover, biotechnology is an ever-advancing field. The progress of this field in the last hundred years, for better or for worse, was enormous. Animal cloning, human *in vitro* fertilisation and the manipulation of the embryo's genetic material, the generation a wide range of vaccines and of induced pluripotent stem cells, as well as the molecular study of a range of diseases. Perhaps in the next decades new technologies will come into being that could generate such an artificial pre-zygote[10] using seemingly licit materials and techniques where it could be used for important research. Or, perhaps, a new treatment will be developed that could be applied in utero at the pre-zygotic stage to prevent a disease by affecting only future somatic cells. Moreover, knowing what event makes us become human might also become relevant in the advent of human augmentations – something that is also slowly encroaching on society.

St John Paul II tells us that life deserves utmost respect and even the

---

[9] Velleman JD. "Family History". *Philosophical Papers*, 2005; 34:357–78 & Mackellar C. "Representative Aspects of Some Synthetic Gametes". *The New Bioethics*, 2015.

[10] Sometimes referred to as the ootid, though the term ootid is also applied to the ovum at a specific developmental stage. Leone S. "The Pre-zygote Identity as a Moral Issue". *Human Reproduction & Genetic Ethics*, 2008; 14:15–21 & García-Rodríguez A, Gosálvez J, Agarwal A, Roy R, Johnston S. "DNA Damage and Repair in Human Reproductive Cells". Int J Mol Sci 20:31.

potential of there being life warrants caution with how we progress.[11] But legislators will demand precision, and only with precision will we be able to rationalise our stance. We need to know when we are dealing with an ensouled human beings so that we can defend them, and to be able to assess emerging technologies. Bioethics here cannot lag behind technology; bioethics needs to be a step ahead. Yet, we also need to know when it is licit to act, so that we do not senselessly oppose beneficial interventions – for respect for life does also entail acting for its benefit. We should not be paralysed by ignorance, but rather guided by sound doctrine and thorough scientific investigations – *fides et ratio*.

Mary as a Guiding Light

Mary plays a special role in our salvation. She guides us in her motherly care and undoes the knots present in our lives. She is not only the archetype of the perfect disciple, but as from Scripture we learn aspects of Divine Law that we might not be able to simply reason from the first principles of Natural Law, so from Mary's life we can learn something about ourselves that we cannot discover from science or philosophy. As this chapter will show, her conception can guide us to important answers regarding our own beginning, and so prevent us from entangling ourselves in the traps of human experimentation.

---

[11] *Evangelium Vitae*, para 60.

CHAPTER FIVE

THE FIRST INSTANT OF MARY'S ENSOULMENT[1]

General Introduction to Chapter Five: Gratitude and Ingratitude. There is, as it were, a response of gratitude to human existence: a sense of gratitude which recognizes that being in existence is a gift, whole and entire; indeed, in the words of Hans Urs von Balthasar, gratitude is twofold: 'It is essential, however, to bear in mind that the two trajectories of thanksgiving – towards one's parents and toward God – are not in competition; they exist next to, together with, and in each other'[2]. This gratitude, moreover, is essentially human: 'Therefore, insofar as he is "born of woman", Christ owes thanks for himself to his Mother, and he has to do so in order to be man in the full sense'[3]. In 'Mary, the Church is embodied even before being organized in Peter'; and, therefore, 'The Church is primarily feminine

---

[1] This was first drafted in 2006-2007, although the General Introduction is wholly new as are a number of discreet parts. Cf. also other accounts of this discussion e.g. *Scripture: A Unique Word*, Newcastle upon Tyne: Cambridge Scholars Publishing, 2014, pp. 311-317 and *Volume III- Faith is Married Reason*, also published by Cambridge Scholars Publishing, 2016, p. 98. I also owe a general debt to the Rev. Dr. Richard Conrad OP for many helpful discussions at a particularly formative time of my interest in questions concerning the beginning of each one of us; and, as noted in this chapter, he is still a helper in the work which Providence still makes possible. A version of this chapter is published in the *National Catholic Bioethical Quarterly* of America (Autumn, 2019).

[2] Hans Urs von Balthasar, "The Marian Mold of the Church", p. 127 of *Mary: The Church at the Source*, 2005.

[3] *Ibid*, p. 130.

because her primary, all-encompassing truth is her ontological gratitude, which both receives the gift [of salvation] and passes it on'[4]. In other words we are all caught up in a movement of gratitude that begins in being given existence as a gift and grows as it is permeated by the grace of salvation which comes through Christ and His Church.

By contrast, then, there is an attitude which expresses a desire 'to owe no one thanks for himself' and, indeed, to react to man's 'vexing dependence as an alienating bondage' to the extent of claiming that 'man is more himself the less he gets his being from others'[5]. Thus there is the possibility that the deepest part of the pain of being imperfectly human is, as it were, an ingratitude which expresses the rejection of the *being-gift* of human personhood. It therefore seems possible that the problem that manifests itself in a variety of ways, particularly in the bioethical crises that confront us, is the problem of an existential ingratitude: an ingratitude that expresses a fundamental revolt against the gift of being; however, rejecting the gift of being does not remedy the problem of being who I am but, rather, constitutes an almost insurmountable obstacle to the reconciliation that brings peace: the reconciliation between the person and the reception, as it were, of human existence from God and from our parents.

Drawing on these initial thoughts, then, it can be said that a discussion on the dogma of the *Immaculate Conception* is a discussion which both ontologically roots Mary's response of gratitude to God and, at the same time, reveals essential elements to a modern discussion of bioethical questions. In other words, both because Mary is a true human being, created as we are, and because she has been the subject of a dogmatic pronouncement about her original sinlessness, she is the perfect exponent, as it were, of our own beginning. There is the view, however, that 'The phrase, [preserved free from all stain of original sin] "in the first instant of her Conception," avoided

---

[4] *Ibid*, p. 140.

[5] *Ibid*, p. 126.

debates which belong more properly to biology'[6]. Nevertheless, as it was 'The person of Mary, not merely her soul, [that] was the subject of the immunity from original sin'[7], it follows that it is possible to explore this dogma in the hope of it indirectly illuminating the question of human conception of the first instant of Mary's conception; for the unity of the human person, one in body and soul (*Gaudium et Spes*, 14), implies the possibility that the moment of Mary's conception being "uniquely required" by the 'singular grace and privilege granted by Almighty God, in view of the merits of Jesus Christ, the Saviour of the human race'[8] is the first instant of fertilization. In view, however, of the possibility that Pope Pius IX did not have the first instant of conception in mind when he promulgated *Ineffabilis Deus*, he says that his predecessors 'Definitely and clearly … taught that the feast [of the *Immaculate Conception*] was held in honor of the conception of the Virgin'[9].

Chapter Five: Introduction: Two Possible Meanings of the One Reality of Human Conception. The modern documents of the Church advert to the following philosophical question: Is the human soul created by God at conception or at some subsequent moment such as that of nidation, otherwise known as implantation?[10]

The argument advanced here is that the Church does in fact make the following affirmation: the soul is one with the body from conception – but this single answer appears to have two possible interpretations. On the one

---

[6] Adapted quotation from "Immaculate Conception, The", p. 179 of Michael O'Carroll's, *Theotokos: A Theological Encyclopedia of the Blessed Virgin Mary*, Eugene, Oregon: Wipf and Stock Publishers, 2000.

[7] O'Carroll, "Immaculate Conception", p. 179.

[8] O'Carroll quoting from 'The Papal Bull, *Ineffabilis Deus*', p. 179.

[9] *Ineffabilis Deus*: http://www.newadvent.org/library/docs_pi09id.htm.

[10] Cf. *Declaration by the Sacred Congregation for the Doctrine of the Faith on Procured Abortion*, footnote 19, p. 16 and cf. also *Donum Vitae*, I, 1 and *Evangelium Vitae*, art 60.

hand the *Dogma of the Immaculate Conception* implies that there is a first instant of fertilization and that this is *necessarily prior to the first instant of the fusion of the male and female gametes*. The reason that the dogma could be said to imply a first instant of fertilization is this: that if Mary is wholly free from original sin, one in body and soul, then this freedom existed from the earliest possible moment of her existence: the first instant of her existence. On the other hand there is the possibility, particularly in the English translation of *Donum Vitae*, of a definition of conception which is subsequent to the *very first instant of fertilization*[11]. It says in the English translation of *Donum Vitae*: 'Thus the fruit of human generation, from the first moment of its existence, that is to say from the moment the zygote has formed, demands the unconditional respect that is morally due to the human being in his bodily and spiritual totality'[12]. The key point, here however, arises from a note that does not occur in the Latin text, namely, that '*The zygote is the cell produced when the nuclei of the two gametes have fused'[13]. Thus there is the impression that the note is interpreting the document to mean that 'the moment the zygote has formed' is when 'the nuclei of the two gametes have fused'; and, therefore, this text appears to advance the view of a slightly delayed moment of human conception. In view, however, of the Latin text of *Donum Vitae* omitting the sentence that defines the zygote means that the interpretation of the Church's understanding of conception remains open, as it were, and could therefore be understood to mean that conception is from the first instant of fertilization: a moment prior to the

---

[11] There are more extensive discussions of different translations of *Donum Vitae* in two other works: *The Human Person: A Bioethical Word* and *Conception: An Icon of the Beginning* (both published by En Route Books and Media).

[12] *Donum Vitae*, I.1.

[13] This note comes at the end of section I.1: "What Respect is Due to the Human Embryo":

http://www.vatican.va/roman_curia/congregations/cfaith/documents/rc_con_cfaith_doc_19870222_respect-for-human-life_en.html.

fusion of the nuclei of the egg and the sperm. The presumption here is that what exists from the first instant of fertilization is a whole and not just parts; and, therefore, presuming that there is no obstacle to the whole being present, this is the beginning of the whole human being. What constitutes the whole human being, and what therefore is present at the first instant of fertilization, is what continues to be discussed in the further development of this essay.

This discrepancy between the very first instant of Mary's conception, one in body and soul, as understood in the Dogma of the *Immaculate Conception* and a possible interpretation of *Donum Vitae* that there is a subsequent moment of human conception, namely that of the fusion of the nuclei of sperm and egg, leads to the view that there is still progress to be made in the Church's understanding of the *mystery of the originating moment of human personhood*. I therefore hope that these reflections, inadequate as they are in some respects, will nevertheless contribute to the recent debate on the issues surrounding the beginning of the human life and which raise, therefore, more or less directly, the very question of the *beginning of each one of us*[14].

1. The Dogma of the *Immaculate Conception*: Two Instants or One Moment?

While it did not seem to be even a secondary intention of Pius IX, while defining the dogma of the *Immaculate Conception* in *Ineffabilis Deus*, to resolve the question of the moment of the animation of the body by the soul, yet the formulation of this dogma required a *precise sense* to be given to such a moment if that same moment was to be the moment of the sanctification

---

[14] I am particularly thinking of the Altered Nuclear Transfer debate in previous issues of *Communio* and to what looks like a wide-ranging call from David L. Schindler for contributions to this debate, "Veritatis Splendor *and the Foundation of Bioethics*", Communio, Vol. 32, no.1, (Spring 2005): 201: 'Christians ... no longer have the luxury of leaving such ponderings to "specialists".'

of Mary, the Mother of our Lord and Saviour Jesus Christ. What, then, is the moment of Mary's sanctification so defined; and does the definition of that moment have an implication of benefit to this inquiry?

In the first place the conception of Mary is venerated as something extraordinary[15]; however, what is extraordinary is the moment[16] and degree[17] of her *sanctification*, not the divine-human *act of her conception*. Secondly, 'The Roman Doctrine', as it is called, developed in part in answer to the possibility of a distinction between a first and a second instant of Mary's conception. And while these instants are not categorically defined, yet there seems sufficient in the text to suppose that the first instant was the *absolutely first instant of Mary's conception* and the second instant was the moment of her sanctification[18]. Thus it could be argued that Alexander VII, in 1661[19], replied to the possibility of a difference between a first and a second instant of Mary's conception and sanctification in the following way: the devotion of the faithful is based 'on the belief that her soul, in the first instant of its creation and in the first instant of the soul's infusion into the body, was ... preserved free from all stain of original sin (...)"[20]. What, then, are we to understand by these two instants?

Is there, in other words, an implicit answer to the moment of human conception in the Dogma of the Immaculate Conception of Mary?

A question that arises, then, is whether it is possible to conclude from this statement that there is an implicit doctrine that the first instant of the existence of Mary's soul, and the first instant of the infusion of that soul into the body, are one and the same instant?

---

[15] Pope Pius IX, *Ineffabilis Deus*, (Boston: St. Paul Books & Media), p. 5.

[16] Cf. *Ineffabilis Deus*, p. 11.

[17] Cf. *Ineffabilis Deus*, p. 17.

[18] *Ineffabilis Deus*, p. 7.

[19] *Ineffabilis Deus*, footnote 7, p. 24.

[20] *Ineffabilis Deus*, pp. 7-8.

The openness of the wording of the dogma of the Immaculate Conception

To begin with the way that the dogma is worded does not preclude the possibility that these two instants are in fact one and the same instant: that the instant of the soul's creation and the instant of ensoulment are one and the same instant. Therefore it is possible that these two 'instants' are in fact one instant, but considered as two from the point of view of the two aspects of the mystery under consideration: the soul's creation and its creation in union with the body. What the dogma of the *Immaculate Conception* affirms, then, is not that the body was in existence before the soul, but that the soul did not pre-exist the body. This is also what would follow if a soul is by definition *the life of the body*[21]. How could the soul exist as prior to the body whose life it is? Or how could a being exist without the 'form' which determines what it will be? What if body and soul reciprocally condition each other and, therefore, require what will constitute the whole to be present on conception? Therefore the soul must come into existence either at the moment at which the body as a body comes into existence or at some subsequent point when the matter ceases to be matter, however animated, and becomes a human body in virtue of being rationally ensouled.

The language of this affirmation concerning the first instant of the sanctification of Mary's soul, is a language which directs itself to the soul of Mary and not to the person of Mary, one in body and soul. Thus Alexander VII says: "ancient indeed is the devotion of the faithful based on the belief that her soul ... was ... preserved free from all stain of original sin"[22]. In other words, it could be argued, the very terms in which Alexander VII expresses himself are not as it were an answer to the question of the precise *time* at which God gives the soul to animate the body.

---

[21] *Catechism of the Catholic Church*, CCC, 365: 'it is because of its spiritual soul that the body made of matter becomes a living, human body'.

[22] *Ineffabilis Deus*, p. 8.

On the one hand, if Mary's soul is sanctified at the first instant of existing, then it follows that this is the first instant of the soul being enfleshed, as it were, in the body; for, in reality, the soul is the soul of a particular body and not an "abstract" entity awaiting a "body" – just as the body is not a human body until the moment it is ensouled. As, then, the sanctification of Mary's soul from the first instant of its creation is asserted, it follows that these two 'instants' are one and the same 'instant'.

On the other hand, in the section of *Ineffabilis Deus* entitled 'Explicit Affirmations', Pius IX says that the Fathers 'affirmed that the Blessed Virgin was, through grace, entirely free from every stain of sin, and from all corruption of body, soul and mind'[23]; and then afterwards, in the section entitled 'Of a Supereminent Sanctity', he goes on to say: 'They testified, too, that the flesh of the Virgin, although derived from Adam, did not contract the stains of Adam ...'[24].

Does the fall of creation on the sin of Adam and Eve contribute to understanding the moment of Mary's Immaculate Conception?

In order, however, to explore this question it is necessary to propose a kind of thought experiment based on the ancient understanding that there was a succession of souls, namely plant and animal, before the rational ensoulment which then brought about the presence of the whole human being; indeed, St. Thomas Aquinas furnishes a helpful understanding of this possibility when he says: 'foetuses are animal before they are human ... [but] nature, in producing the animal foetus, is aiming at producing a man'[25]. The purpose of this thought experiment is to make more explicit the necessity of

---

[23] *Ineffabilis Deus*, p. 16.

[24] *Ineffabilis Deus*, pp. 16-17.

[25] St. Thomas Aquinas, *Summa Theologiae*, A Concise Translation. Edited by Timothy McDermott, London: Methuen, 1989: I, 85, 4.

the presence of the whole human being from the first instant of fertilization; for, if Mary is wholly sanctified from the earliest possible moment of human conception, and Mary shares our human condition in all things except sin, then it follows that the Dogma of the *Immaculate Conception* also defines the first moment of human conception. Thus there arises the following question: If the body-to-be of Mary is conceived *before* the coming into it of the rational soul, then would that precursor human body be free of its participation in the fall of creation?

For, in general, if 'Because of [the fall of] man, creation is now subject "to its bondage to decay"'[26], then it follows that the matter and indeed all that is entailed in the formation of a body is 'subject "to its bondage to decay"'. In other words, if Mary is to be *absolutely free, body and soul,* of both the subjection of creation to futility (cf. Rom 8: 20) and decay (cf. Rom 8: 21) *and* original sin which is "the death of the soul"[27], then it would follow that the moment that her soul was created, sanctified and infused *was* the first instant of fertilization in that this was the first possible moment that she could have come to exist, one in body and soul, beneath the heart of her mother Anna[28]. Thus if one had to choose between the following two possibilities: either that of the *immediate animation of Mary, one in body and soul,* or the *belated* animation of the precursor-body by a rational soul, it would seem that the one which would yield the *perfect conception of Mary*, is the one which would allow no interval of time between the sanctification of Mary's soul and the creation of her body.

Finally, then, one is driven by argument to the view that Mary was conceived and sanctified, body and soul, at conception, and this is precisely what could be said to be implicit in the words by which Pius IX defines the

---

[26] CCC, 401, Rom 8: 21.

[27] Cf. Council of Trent: DS 1512, as quoted in the CCC, art 403.

[28] Cf. *Ineffabilis Deus*, p. 16; but the expression, 'beneath the heart' comes from elsewhere, possibly from the writings of Pope St. John Paul II.

Immaculate Conception : 'Mary, in the first instant of her conception ... was preserved free from all stain of original sin ...'[29]. For Pius IX does not say that in the *first instant of the existence of Mary's soul*, that it was preserved free from all stain of original sin; rather, he says: 'Mary, in the first instant of her conception ... was preserved free from all stain of original sin ... '., where conception is now *understood* to be the *absolute beginning of Mary, one in body and soul*.

2. The Help of the Doctrine of Original Sin

Given that the body of Mary cannot exist until the moment of ensoulment it is necessary to use an expression, however inadequate, to describe the result of the thought experiment, namely, that prior to Mary coming to exist there could have been a "precursor body": a "body" that nature intended to be Mary's. The reason that Mary's "precursor body" could not have been conceived *before the moment of the creation and infusion of her soul* is that in a general way her "precursor body" would have been subject to the condition of the fall of creation. On the basis of such a chronological order of things, the "precursor-body" of Mary would in some way require the rectification of the imperfection it inherits through the *fallen matter* of which it is made. This rectification would therefore have had to come about through the precursor body's 'share' in the sanctification of Mary's soul *on the moment of its conception and infusion*. While it is possible, however, for God to intervene and to prepare the "precursor body" for the reception of Mary's soul it is more coherent for there to be a single moment of human conception in which Mary is immaculately conceived; and, indeed, it is increasingly clear that there is in reality a single moment of human conception in that the life of the soul is the life of body: there is a kind of incarnation of the soul "in" the body (cf. *Familiaris Consortio*, 11). Therefore, *if Mary's body was to share*

---

[29] *Ineffabilis Deus*, "The Definition", p. 21.

in the gift of sanctification given to the soul immediate to its coming into existence in such a way that her body was *never* to have been in any kind of fallen state[30], then the body of Mary was itself to have come into existence at the same moment that it expressed[31] the sanctification of Mary's soul. Thus it could be said that Mary was sanctified, body and soul, at conception: at a conception in which body and soul are wholly one and one from the very first instant of her existence.

The help of original sin: sin is transmitted through the flesh

The doctrine of original sin, however, offers a more precise understanding of how one can reject the possibility that Mary's "precursor-body" *could have existed prior to the creation, sanctification and infusion of her soul.*

St. Augustine tended to the view, it seems, that the soul cannot be immediately created by God. For 'If the soul came straight from God, how could it come stained with original sin?'[32]. In answer to this difficulty St. Thomas Aquinas 'places the essence of original sin in" the privation of

---

[30] Cf. CCC, art 404.

[31] Cf. for instance the following: St. Thomas Aquinas says of the body and soul at the resurrection of the body: 'our body will be united to our pre-existing soul, but not as something secondary supervening, but as something taken up into the same existence, receiving life from the soul' (Pt III, Qu 2, art 6); and the 'body shares one natural existence with the soul ...' (Pt III, Qu 2, art 6). The salient point is that the body cannot exist before the sanctification of the soul, if the sanctification of the body is to share *in the same moment of the sanctification of the soul.* These excerpts are from the *Summa Theologiae*, a concise translation by T. McDermott, (London: Methuen, reprinted 1992).

[32] *A Catholic Dictionary*, by William E. Addis and Thomas A. Arnold and revised with additions by T. B. Scannell, (London: Virtue & Company Ltd, ninth edition, 1916), p. 783: Soul; hereafter *Catholic Dictionary*.

original justice" ... '[33]. Now the author of this article from which I have just taken a second quote goes on to say that it is this concept of original sin as a privation that comes to our help. For God 'can and does infuse souls deprived of original justice ...'[34].

In 1546 the Fathers of the General Council of Trent wrote a *Decree on Original Sin*. In this decree they express the following characteristics of original sin: it is 'one in origin' ; it is 'transmitted by propagation'; it is 'in all men, proper to each; and it cannot be taken away by any 'remedy other than the merits of the one mediator our Lord Jesus Christ who reconciled us with God by His blood, being "made our righteousness and sanctification and redemption" (1 Cor. 1: 30)'[35]. In the following article they go on to say that the baptism of children is therefore necessary 'so that by regeneration they may be cleansed from what they contracted through generation. For "unless one is born of water and the Spirit, he cannot enter the kingdom of God" (Jn. 3: 5)'[36].

What is contracted by generation?

What, then, can be contracted by generation? What the parents transmit *through the moment of fertilization and ensoulment is the beginning of a person's life in a state which is deprived of its original perfection as given by God to Adam and Eve.* In other words, *precisely because* Adam and Eve are our first parents what they lost of what was originally given to them is, as it were, *personally transmitted to us through the act of human generation.* This

---

[33] *Catholic Dictionary*, p. 633: Original Sin.

[34] *Catholic Dictionary*, p. 634: Original Sin.

[35] *The General Council of Trent Fifth Session Decree on Original Sin*, art 1513, as in *The Christian Faith*, edited by J. Neuner SJ and J Dupuis SJ, (New York: Alba House, revised edition 1982); hereafter *The Christian Faith*.

[36] *The General Council of Trent Fifth Session Decree on Original Sin*, art 1514, *The Christian Faith*.

would imply that the original grace which Adam and Eve lost was an original grace which *inhered, permeated or otherwise included* their bodily substance.

This also accords with what St. Thomas Aquinas says of grace: 'grace is of a higher order of reality than the soul, but not in its mode of existing'[37]. The mode of existing of grace is that of a *supervening quality* which, by definition, is not so much something that exists in itself as the way in which something else exists: 'and so grace is not created, but men are created in it, established in a new existence out of nothing, without earning it: *Created in Christ Jesus in good works*'[38]. In other words, the loss that is transmitted through the 'Fall' in general and the generation of the flesh particularly, is the loss of what is created *being created to exist in the grace of God*. Therefore, whether *from the point of view of the transmission of the flesh or the creation of the individual soul by God on conception,* in either case the new person comes into existence *deprived of the Creator's original fullness of gift: which was to be created in the grace of God.*

The relevance of all this to the discussion on Mary is that *therefore* Mary must have been sanctified on conception, one in body and soul, *or her body would have been contaminated by original sin, as understood to be transmitted as a deprivation through human generation*. Finally, because Mary is a creature like ourselves in all things except this contamination of original sin, then the fact of the existence of the unity of body and soul from conception, presupposed, it would seem, by the doctrine of her *Immaculate Conception*, is a fact that is common to the beginning of human beings.

An Objection: Cannot grace be given by God at any time?

Why could not a grace be given to the body in such a way that while it came into existence before the soul it did not come into existence 'with' the

---

[37] *Summa Theologiae*, Methuen, Pt II, Qu 110, art 2.
[38] *Summa Theologiae*, Methuen, Pt II, Qu 110, art 2.

privation of grace that we call original sin? For if 'grace is not created, but men are created in it, established in a new existence out of nothing, without earning it: *Created in Christ Jesus in good works*'[39], then it does seem possible to say that the body of Mary could have been created by God *in such a way as it would have been without the privation of original sin.*

Now we accept that creation was made *through* Christ (Jn. 1: 3). And, therefore, could something be made through Christ without receiving the benefit of the 'fulness' of grace of which He is the bearer (Jn. 1: 17)? Therefore, if creation itself participated in the grace of God *through being made in Jesus Christ*, then there is no particular obstacle to the body of Mary being made in a similar way *in Christ.*

The problem, however, is that Mary was not conceived in Christ at the moment that creation was made through Christ. In other words, her body did not exist from creation in some hidden form until animated by her soul. At the time of Mary's parents conceiving her, then, while there is the possibility that there was a kind of "precursor-body" which was exceptionally preserved from sin, there is also a more coherent explanation of how Mary was preserved from original sin, one in body and soul: an explanation that draws on the wholeness of human personhood and the mystery of the grace of God.

Grace[40], according to St. Thomas Aquinas, requires the existence of a mind: 'it is through the mind that man ... receives grace'[41]. The existence of a mind, even to the degree which corresponds with this first moment of

---

[39] *Summa Theologiae*, Methuen, Pt II, Qu 110, art 2.

[40] Following the Rev. Dr. R. Conrad's, OP, comment (4/07/01) and conversation on the (13/07/01), there are two complex questions to be addressed: the first is the nature and effect of grace or graces (sanctifying grace and others); and the second are the reasons for rejecting other anthropologies. But it is beyond the scope of this particular book to pursue these points here.

[41] Cf. St. Thomas Aquinas, *Summa Theologiae*, Methuen, III, 5, 4: 'it is through the mind that man sins and receives grace ... '.

existence[42], implies the existence of a soul. If there is no soul, then there is no possibility of a union with Christ to prevent the infection of the fall; and if there is no soul then there is no possibility of a grace integrating body and soul[43]. Grace, then, to summarise St. Thomas, requires a personal "subject" as it is a benefit to the person; and, therefore, grace requires the presence of the soul[44]. In the words of St. John Henry Newman, grace is 'a real inward condition or superadded quality of the soul'[45]; and, drawing on an Anglican divine, Newman quotes exactly the point to be made with respect to the

---

[42] Cf. John Paul II, *Evangelium Vitae*, 60.

[43] CCC, 400.

[44] Other references to the following work were suggested by the Rev. Dr. Richard Conrad, OP, (email, 10/3/2019): Cf. St. Thomas Aquinas, *Summa Theologiae*, Methuen, 1992: I-II, Qu 113, Art 10: 'the soul has a *natural capacity for grace being made in the God's image*' (p. 321); I-II, Qu 63, Art 2: If the standard of virtue 'is God's law then ... [it] can only be caused by an activity of God within us' …. [furthermore, taking account of the possibility of the presence of original sin, it is possible to say with St. Thomas that 'Such divinely instilled virtue cannot co-exist with mortal or fatal sin' (p. 241); I-II, Qu 5, Art 5: 'a nature that can thus achieve utmost perfection, even though needing external help to do it, is of a nobler constitution than a nature that can only achieve some lesser good, even though without external help' (p. 181); I-II, Qu 1, Art 8: 'Men attain their goal by coming to know God and love him' (p. 174); I, Qu 93, Art 4: 'grace adds to some men an actual if imperfect understanding and love of God' (p. 144) which, in the case of the Blessed Virgin Mary, goes to the limit of human perfection and excludes even the possibility of original and personal sin; I, Qu 8, p. Art 3: 'God exists in those actually knowing and loving him, or disposed to do so; and since this is God's gracious gift to reasoning creatures we call it existing by *grace* in his chosen friends' (p. 22).

[45] The quotation is cited from *Difficulties Felt by Anglicans*, Vol. II, p. 46, published on pp. 22-23 of *Mary: The Virgin Mary in the Life and Writings of John Henry Newman*, edited with an Introduction and Notes by Philip Boyce, Leominster, Herefordshire: Gracewing Publishing, 2001.

presence of Mary's soul at the moment of the reception of grace: 'Adam was created in grace, that is, received a principle of grace and divine life from his very creation, or in the moment of the infusion of his soul'[46]. In other words, the presence of the soul for the reception of grace seems an ecumenically established truth. If grace requires the presence of the human soul, then for grace to be effective in the flesh, as it were, as well, then body and soul need to be united. Thus the mystery of the *Immaculate Conception* implies that Mary is one in body and soul (*Gaudium et Spes*, 14) at the instant of their reciprocally coming to exist; indeed, as it says simply in *Lumen Gentium*: 'Enriched from the first instant of her conception with the splendor of an entirely unique holiness, the virgin of Nazareth is hailed by the heralding angel, by divine command, as "full of grace" (cf. Lk. 1: 28 …)' (56). In other words, while the Church does not explain the 'first instant of conception' – the ultimate 'first instant' is the first instant that the sperm animates the egg and the embryo expresses this through the formation of the embryonic wall[47].

A confirming consideration from St. John Henry Newman

Newman seems to recognize the problem under consideration when he contemplates what is necessary if Mary, the Blessed Virgin and spouse of St. Joseph, is to conceive and bear the Son of God in her womb:

'Who can estimate the holiness and perfection of her, who was chosen to be the Mother of Christ? …. [W]hat must have been the transcendent purity of her, whom the Creator Spirit condescended to overshadow with His miraculous presence? … This contemplation runs to a higher subject,

---

[46] The Anglican divine from whom Newman is quoting is Bishop George Bull (1634-1710), on p. 224 of *Mary: The Virgin Mary in the Life and Writings of John Henry Newman*.

[47] Cf. Chapter 12, Etheredge, *Scripture: A Unique Word*, 2014.

did we dare follow it; for what, think you, was the sanctified state of that human nature, of which God formed His sinless Son; knowing as we do, "that which is born of the flesh is flesh," and that "none can bring a clean thing out of an unclean?"'[48].

An additional consideration that arises, then, is through Newman's echoing of the beginning of creation which is itself echoed in St. Luke's Gospel. On the one hand it says in St. Luke's Gospel: 'And the angel said to her, "The Holy Spirit will come upon you, and the power of the Most High will overshadow you; therefore the child to be born will be called holy, the Son of God"' (Lk. 1: 35). On the other hand it says in Genesis: 'The earth was without form and void, and darkness was upon the face of the deep; and the Spirit of God was moving over the face of the waters' (Gn. 1: 2). In other words, just as there is a general relationship between the original beginning of creation and the conception of Jesus Christ, just as there is the presence of the Blessed Trinity[49] in the beginning and at the *Incarnation*[50], so just as creation came from the Creator blessed and unblemished so the unblemished flesh of the 'Word' (Jn. 1: 14) came from the unblemished flesh of the Blessed Virgin Mary.

We can, with Newman, sum up by saying 'that grace was given ... [to Mary] from the first moment of her existence' and that 'He gives grace and regeneration at a *point* in their earthly existence; [but] to her, from the very beginning'[51]. Thus, although St. Thomas Aquinas could not resolve the

---

[48] The quotation is cited from *Parochial and Plain Sermons*, Vol. II, pp. 131-2, published on pp. 22-23 of *Mary: The Virgin Mary in the Life and Writings of John Henry Newman*.

[49] Cf. Etheredge, *Scripture: A Unique Word*, "Chapter Eight: Creation: The Archetypal Action of God", p. 252.

[50] Cf. Balthasar, *Mary for Today*, p. 35.

[51] Excerpts from "A Memorandum on the Immaculate Conception", 3 and 4 on p. 304 of *Mary: The Virgin Mary in the Life and Writings of John Henry Newman*.

difficulty entailed in it being true that Mary was conceived without sin[52] yet, as we have seen, the wealth of his arguments is ever useful even if the following use of it is adapted to a new purpose; and, therefore, if the *Immaculate Conception* is indeed a new beginning which echoes the originality of creation, as it were, God begetting anew the beginning of salvation, then it is fitting[53] that Mary is conceived as *per* the beginning of creation: So Mary's 'body was formed by God immediately'[54] He ensouled her.

3. A Discrepancy Between the Implied First Instant of Fertilization in the Dogma of the *Immaculate Conception* and the Definition of Conception in the English Translation of *Donum Vitae*

The argument is, then, that the Dogma of the *Immaculate Conception*, in conjunction with the doctrine of *Original Sin*, imply a holistic account of the first instant of human conception. This is because the manifest objective of *Ineffabilis Deus* was not the doctrine of our beginning but the moment of Mary's *Immaculate Conception*: a moment that implied an answer to the question of when she began. Reflecting on the moment of the conception of

---

[52] 'It seemed to ... [St. Thomas] that if ... [Mary] had been preserved from original sin, the absolute and universal need of redemption would no longer be valid (cf. *In IV Sent.*, d. 43, q. 1, a. 4, s. 1 and 3; *Summa Theologiae*, III, q. 27, a. 2)' (footnote 5, p. 306 of *Mary: The Virgin Mary in the Life and Writings of John Henry Newman*). The objection was later overcome by Blessed Duns Scotus who argued that 'Mary needed to be redeemed and in fact was redeemed, but in a unique way – by anticipation' (footnote 7 on p. 229 of *Mary: The Virgin Mary in the Life and Writings of John Henry Newman*.

[53] "Fittingness" is an argument that pervades this whole question and was used by Newman (cf. pp. 75-77 of *Mary: The Virgin Mary in the Life and Writings of John Henry Newman*).

[54] *Summa Theologiae*, Methuen, p. 142: I, Qu 91, Art 2.

Mary, one in body and soul, as the moment of the first instant of Mary's existence and reception of grace nevertheless gives new impetus to the question of the first instant of human personhood. Indeed, while there is a kind of dynamic in the person of Christ and of His mother Mary, it is not until the descent of the Holy Spirit at Pentecost that the Holy Spirit comes to dwell in the Church like the soul in the body (*Lumen Gentium*, 7); and, therefore, the unity of the Church does not exist prior to Pentecost nor does conception occur prior to the soul's creation and ensoulment in the first instant of fertilization. Nevertheless, because specifying the moment of our beginning *is a necessary presupposition* to the specification of the dogma of Mary's *Immaculate Conception*, it has been argued that the Tradition of the Church contains an implicit affirmation of a doctrine of the moment of our beginning.

This *implicit affirmation* begins to be explicit in the modern documents of the Church; however, as already indicated, when this implicit affirmation becomes explicit there is, it seems, an element of contradiction between two affirmations. On the one hand, it has been argued that the Dogma of the *Immaculate Conception necessarily affirms* a first instant to the existence of Mary that is prior to the fusion of the nuclei of the sperm and the egg or ovum. In other words, in the language of the document on the Church, there is a simple 'first instant' of Mary's existence: 'Enriched from the first instant of her conception with the splendor of an entirely unique holiness, the virgin of Nazareth is hailed by the heralding angel, by divine command, as "full of grace" (cf. Lk. 1: 28 …)' (*Lumen Gentium*, 56).

This first instant of Mary's existence, in modern terms, is *necessarily the first instant that the sperm animates the egg*[55]. If this is not the case, then in

---

[55] That there is a first instant of fertilization, is for some a controversial claim and requires some elucidation. Suffice it to say here that the "inertia" of the egg after ovulation is actually transformed by the sperm's penetration of the ovum. This transformation of the ovum's post ovulatory inertia is manifest in the ovum's resultant closing of its pores. In so far as a first instant is either definable or "visible",

modern terms there is a time in which the body of Mary exists and is thus "subject", contrary to the Dogma of the *Immaculate Conception*, to the imperfection of a *necessarily "bodily" kind of matter which is subject to the state of original sin*.

In contrast, then, to the aforementioned implication of the Dogma of the *Immaculate Conception*, there is the affirmation of the *Congregation for the Doctrine of the Faith*, which says: 'the fruit of human generation, from the first moment of its existence, that is to say from the moment the zygote has formed, demands the unconditional respect that is morally due to the human being in his bodily and spiritual totality'[56]. In the English text at least, then, there is the possible interpretation of a characteristically delayed moment of human animation in that 'The zygote is the cell produced when the nuclei of the two gametes have fused' [57].

Conclusion: The Need for a More Precise Definition of Human Conception in the Documents of the Church

In a word, while there is no question of the foregoing discussion having a negative impact on the treatment of the person at the moment that each one of us comes to exist[58], nevertheless it seems clear that either this understanding of the Dogma of the *Immaculate Conception* is erroneous or there is work to be done concerning the further doctrinal exposition of the meaning of conception: of the meaning of the moment that each one of us

---

the sperm's stimulation of the ovum and the ovum's response to it is the *first instant of fertilization*. Cf. *A Child Is Born*, photographs by Lennart Nilsson and text by Lars Hamberger, translated by Clare James, (London: Doubleday [Transworld Publishers Ltd.], 1990), p. 51.

[56] *Donum Vitae*, I, 1.

[57] *Ibid*.

[58] *Evangelium Vitae*, art. 60, second paragraph.

comes to exist in the integrity of being one in body and soul[59]. At least one of the clarifications that is necessary is defining the first instant of fertilization as the beginning of conception: a process that clearly reaches a 'first' stage of development with the formation of the zygote. If the Dogma of the *Immaculate Conception* implies a 'personal presence'[60] from the first instant of fertilization, however, then this defines conception for all of us. For the reasons expressed in this essay, and elsewhere[61], it is clear that there is no better "sign" of the beginning of the person than the *absolutely first instant of the sperm's animation of the egg: an outward sign of the inward action of God*.

Furthermore, if the argument of this chapter is correct then it follows that Mary, one in body and soul, was Mary from the first instant of her conception and not at some subsequent point because it was in virtue of who she was from conception, a particular woman, that the grace of being conceived without original sin was integral to her conception[62]. Therefore the integral nature of being a male-person or a female-person is from the first instant of conception even if, in terms of actual human development, there are reasons why this is not without its developmental difficulties[63]. On the other hand, however, there is the claim that 'Human foetuses are initially sexually indifferent, possessing primordial structures of both the male and female reproductive systems; they will feminize unless specific signals are

---

[59] Cf. *Gaudium et Spes*, 14.

[60] *Donum Vitae*, I, 1.

[61] Cf. Francis Etheredge: *Conception: An Icon of the Beginning*.

[62] Although the same point applies to Christ, as a male-person from conception, the argument is more difficult to pursue here because He was conceived in a unique way (like Adam and Eve but different again from then in that He had a human mother).

[63] Cf. *The New Bioethics*, 24:2, 176-189, DOI: 10.1080/20502877.2018Nathan K. Gamble & Michal Pruski (2018) "Teleology and Defining Sex",.1468601.

given to them at specific times to masculinize (Mcbride 1977)[64].

The claim, then, that 'Human foetuses' are 'sexually indifferent' must be understood to be a kind of biological expression that "abstracts" biological development from the whole reality of human being and has to be understood, therefore, from the point of view of the totality of the human person as conceived from the first instant of fertilization. The child, then, one in body and soul, is defined by his or her very existence from the first instant of fertilization and, therefore, the whole process of sexual differentiation is an outcome of a beginning which, like all beginnings, determines the stage and timing of what manifests the whole person: the human being-in-relation. In view, then, of actual developmental difficulties and the suffering which is an almost inseparable part of what happens, there is nevertheless a beginning which determines the end; and, therefore, there is an objective reality of the whole person to which all interventions are necessarily to be ordered for his or her total good. If, however, there are still uncertainties that surround these profound difficulties then it is the task of sound and sensitive professionals to pursue and resolve them – both for the sake of the specific individual person and for the benefit of the science of love indispensably expressed in the care of people afflicted with these profound problems.

Finally, if the integrity of human conception begins the totality of human personhood, where twinning that derives from a single human embryo is a subsequent but no less radical beginning for each person, then it follows that this human integrity is decisive for determining the humanity of the human being; and, therefore, whether there are experiments between species or one's which involve a human being's human identity in some other way[65], it is crucial to recall that a person is a human being-in-relation and requires the

---

[64] *The New Bioethics*, 24:2, 176-189, DOI: 10.1080/20502877.2018Nathan K. Gamble & Michal Pruski (2018) "Teleology and Defining Sex",.1468601, p. 178.

[65] Cf. Michal Pruski (2019) "What Demarks the Metamorphosis of Human Individuals to Posthuman Entities?", *The New Bioethics*, 25:1, 3-23, DOI: 10.1080/20502877.2019.1564003.

humanizing help of good relationships – beginning with the intra-uterine relationship to the mother and the father for adequate human development. In other words, justice to the human embryo overrides "so called" experimental procedures which bear no benefit to the child conceived and indeed often entail his or her destruction or unwarranted risk of harm or interference with normal development; and, therefore, justice to the human being entails the right to the completion of the human development that has begun. Thus it is imperative to establish the reality of human beginning as the point of departure for human rights in legislative assemblies throughout the world; and, in so far as it is relevant, that there are no alterations to the sperm or ovum that would profoundly interfere with the expression of a person's humanity. In other words, the reality of each one of us being a gift not only establishes our absolute equality but entails recognizing that the reality of human rights are reciprocal: human rights are rights-in-relationships; and, therefore, a human being is by definition to be guaranteed the conditions that promote adequate human development in virtue of the very fact of being conceived-in-relationship to other human beings – even if this obscured but not altered by unwarranted manipulations of the beginning of human personhood.

FOREWORD TO CHAPTER SIX

DR. MOIRA MCQUEEN

Introduction: Being in Relationship

As we develop from childhood into conscious self-awareness and knowledge of ourselves as individual beings, we discover that we are embedded in a society made up of family, before we even see our place in the larger society. We begin to experience ourselves as individual beings, as created persons distinct from every other creature, with a body, soul and spirit unique in the world. That is an amazing and mysterious discovery, which triggers for many of us a process of questioning 'why things are the way they are,' and many of us conclude that, since there is creation, there has to be a Creator, an originator, a life-giver, a 'maker,' in light of our being 'made.' Christian faith helps nurture a sense of wonder, leading those who heed it to acknowledge God as our Creator and Father. Revelation gives us increased knowledge and appreciation of the Spirit's role in the coming-to-be, death and resurrection of Jesus, while a growing understanding of the significance of Mary's Fiat! expands our sense of the being-in-relationship-to-God of all creatures and helps us recognize the common relationship we have with all other people through and in our creatureliness, all of us made in the Image of God.

When considering theological approaches to bioethics, this recognition helps shape our moral stance on what should, and not just could, be done to aid human flourishing. Thomas Aquinas' view was all-encompassing: besides what might be best for us as individuals, he reminded us that we should take care that the common good also be considered. This prevents us from being overly subjective in situations where others' wellbeing needs to be taken into account and is an important part of the Catholic Church's views on bioethics and other areas of morality: ethical questions, responses, and medical and social factors must be viewed through the broad lens of the common, as well as our individual, good. For example, it is clear that embryonic experimentation is of no benefit to an embryo since it dies in the process, yet at the same time a cure could be effected from that experimentation for our benefit. Catholic teaching insists that even the smallest embryo is to be treated as a person from the time of conception, therefore it must never be treated as a means to an end, but always as an end in itself[1]. Kant also arrived at that conclusion about human persons, (though perhaps not regarding embryos) and Pope St John Paul II describes all persons as the sort of beings towards whom 'the only proper and adequate attitude is love'[2]

Reliance on God as Creator helps us recognize that all life is to be valued and protected from conception until natural death. Regrettably, the fact that biology confirms that life begins at conception has not stopped some from deciding upon a later time before conceding 'personhood,' used juridically for the attribution of rights. Such decisions ensure that embryos have no rights in many countries, and that fetuses have no, or limited, rights. These ethical questions have important consequences for the smallest of human beings, and parents often have difficult decisions to make about, say, the

---

[1] Congregation for the Doctrine of the Faith. *Donum vitae*, 1987, I (5) "...Methods of observation or experimentation which damage or impose grave and disproportionate risks upon embryos obtained *in vitro* are morally illicit for the same reasons. Every human being is to be respected for himself, and cannot be reduced in worth to a pure and simple instrument for the advantage of others. *It is therefore not in conformity with the moral law deliberately to expose to death human embryos obtained 'in vitro'.*"

[2] St Pope John Paul II. *Love and Responsibility* (San Francisco: Ignatius Press) 1993, P. 41.

transfer of an embryo or the continuation of a pregnancy when there are genetic complications.

Although I am a Catholic moral theologian who prays and reflects over bioethical issues while trying to be faithful to Catholic teaching, I admit I have not sought to include Mary in my approach to such questions, at least in 'being-in relationship' with her in any specific way. I do invoke the help of Our Lady of Good Counsel, or Our Lady, Health of the Sick and so on, but writing this introduction challenges me to deepen this aspect in my work. I am already influenced by my own marriage and motherhood, shaped by Catholic values in being open to life (if seven children indicate that!) and to the role of Natural Family Planning as a way for my husband, whose middle name is Joseph!, and myself to express our love without adopting a contraceptive mentality. I have always appreciated the role of Mary as Mother of Jesus, and as Mother of the Church. I tell people that while my name is Gaelic for Mary, the priest who baptized me decided that 'Moira' was a pagan name, so baptized me "Moira Mary." Mary, Mary! Perhaps I needed the double appellation to awaken me to the fact that I do not pay enough attention to the role of the Blessed Mother in my faith, work and family life. The title of this book, "Mary and Bioethics," gives me an opportunity to consider the role of the Blessed Virgin in bioethics a little more, and writing this introduction is helping me open that door.

I. Man: Male and female, Christ and Mary

Once 'formed' as individuals, developed in the knowledge of ourselves as male and female, perhaps educated in the notion of complementarity through St Pope John Paul's views in Theology of the Body and launched into the world of school, work, friendships, most of us begin to experience natural urges towards the other sex and eventually towards marriage and family. The biological experience of hormonal urges that surge when relationships develop is only one part of the story: the male/female personal love relationship is unlike any other relationship, unique to a particular couple in their specific value to each other and irreplaceableness. When such couples are informed by Christian values, they are aware of these experiences as coming from God, in His plan "from the beginning," as Jesus told us. The story of Adam and Eve further explains what this relationship entailed: freedom to be themselves, to discover themselves, to love each other with all freedom, except for one thing, one forbidden fruit: they were not to forget they were created, they were not God. They overstepped and the relationship was broken. In St Pope John Paul II's words, they lost their original innocence and humankind continues to deal with the aftermath, sin, as it seeks to continue to re-establish the kingdom of God[3].

The Catholic Church has a practical, experiential approach to the male/female relationship as well as a deep Scriptural, spiritual and mystical sense of the fullness of the human being. Over the centuries it has perceived that the man-woman relationship, which is fundamental for procreativity, goes beyond that in showing differences between the sexes in outlook and personality, partly shaped by bodily rhythms and hormonal differences and partly by the circumstances in which people find themselves. This is not 'biology as destiny,' but more a recognition that biology plays a large part in the Catholic worldview. Rather than downplaying those bodily aspects, as some feminist or gender theorists do, Catholic anthropology recognizes the human's natural inclination towards partnering, marriage and settling down, along with the need for stability and a willingness to go beyond personal needs and desires to rear and nurture children. This is, in fact, simply what most people do both instinctively and deliberately, as a matter of freedom of will and choice.

---

[3] St Pope John Paul II. "Original Innocence and Man's Historical State." *Catechesis by Pope John Paul II on the Theology of the Body – 18*, General Audience, February 13, 1980.

St Pope John Paul develops the idea of the complementarity of the sexes in a wonderful way in his Theology of the Body. Moved to marry through mutual consent, the couple will find that "The body, which through its own masculinity and femininity helps the two…from the beginning to find themselves in a communion of persons, becomes in a particular way the constitutive element of their union when they become husband and wife"[4]. William May says it is clear from Genesis 2 that the body is personal in nature and that it 'reveals or discloses' the person[5]. Our bodies are integral parts of who we are as persons, not something we can 'change' as we might change our outlook or our clothing. In giving our bodies to each other, we give our very persons, our selves. This is the spousal meaning of the body, called 'self bestowal' in *Gaudium et Spes*, one of the documents of the Second Vatican Council:

> Christ the Lord abundantly blessed this many-faceted love, welling up as it does from the fountain of divine love and structured as it is on the model of His union with His Church. For as God of old made Himself present to His people through a covenant of love and fidelity, so now the Savior of men and the Spouse of the Church comes into the lives of married Christians through the sacrament of matrimony. He abides with them thereafter so that just as He loved the Church and handed Himself over on her behalf, the spouses may love each other with perpetual fidelity through mutual self-bestowal[6].

---

[4] May, William E. *Marriage: The Rock on which the Family is Built*. (San Francisco: Ignatius Press) 1995, p.33.

[5] *Ibid*. p. 35.

[6] Second Vatican Council. "*Gaudium et spes*", "Pastoral Constitution of the Church in the Modern World", in *Vatican Council II: The Conciliar and Post Conciliar Documents*, ed. Austin Flannery (Collegeville, MN: Liturgical Press), 1975, n.48.

A man and a woman joined in the Sacrament of Marriage are a real sign of the love relationship, the Covenant, between Christ and The Church. This is indeed a great mystery!

II. Christ and Mary

The coming of Jesus the Christ as an enfleshed human being was made possible by the obedience of the Virgin Mary to the Father who 'declared' his intention to her. Mary's acceptance was uttered out of obedience to the Father's will, affecting her life from that instant onwards: "Be it done unto me according to thy word!" Her acceptance of and trust in God's will were mirrored in Joseph's acceptance of and trust in the Angel's message to him, and also in his acceptance of and trust in Mary, his betrothed. Surely neither Mary nor Joseph could fully understand all that was happening! Scripture tell us that Mary 'pondered these things in her heart.' She reflected, she prayed, she accepted: the angel acknowledged her as 'full of grace' and clearly, she was an extraordinary person, chosen by God, who in turn totally trusted that God would provide. She bore this son, who was also Son, in the human, spiritually-infused drama described in Scripture, in lowly and foreign surroundings, showing this was no ordinary birth. St Joseph is depicted as being caregiver and guardian of his wife as much as of the baby born to her. All we know of Jesus' earthly father is that he was a carpenter and that the Holy Family lived in Nazareth. There is not a single spoken word attributed to him in Scripture, but as father, he raised Jesus in traditional Jewish practices such as presenting him in the temple and ensuring his circumcision, while being a dedicated worker who provided for his family. St Pope John Paul II and Pope Leo XIII speak of Joseph as 'magnified' by proximity to Mary, and the title of one document, *Redemptoris Custos*, shows the importance of Joseph's role: the guardian of our Redeemer, Jesus the Christ![7]

In their family life at Nazareth, where Jesus lived and worked for thirty years, a Trinitarian model can be seen, exemplifying the first 'domestic church.' What better model could there be for every family that welcomes the Trinity as an active component in their lives: the love of the Father and the Son expressing itself in the Spirit, the mother and father's love expressing itself in their children! Spouses can look to the ongoing work of the Spirit in their salvation, to the model of Mary and Joseph as parents to the model of Jesus and Mary as a bond that gave birth to the Church as Bridegroom and Bride, and to their own 'sign' value in its nuptial significance.

We are told that Jesus showed obedience to Mary and Joseph, although always pointing towards his heavenly Father as His ultimate authority. This knowledge must have reassured Mary and Joseph, even as they occasionally remonstrated: 'Son, why have you done this to us?' All three, Jesus the Christ, Mary and Joseph are models for us to follow in our desire to be obedient to the Father, even when we, too, ask: 'Why have you done this to us?' Parents ask such a question of God in times of our human tragedies: unwanted pregnancies, divorce, early death, addictions, mental illness, other illnesses, suffering (all areas that enter the field of bioethics in one way or another): Why? Why us? I /we don't deserve this, and so on. Acceptance of an unwanted reality is one of the greatest challenges we face and it is interesting how faith sustains some people, how they learn to accept the reality, sometimes only gradually, how they unite their sorrow with Mary's or even with Jesus' Passion, how they ultimately transcend trials and are open to the will of God. Through experience, we know that, often, what seems like a tragedy can work for good. Perhaps during those trials it would help to reflect

---

[7] Dodd, Gloria F. "The Nuptial Meaning of the Body in the Marriage of Mary and Joseph." *The Virgin Mary and Theology of the Body*, Donald H. Calloway (Ed.), (Stockbridge, Mass: Marian Press), 2005, p. 119.

on the burdens faced by Mary and Joseph, and those faced by Mary alone, after the death of Joseph, which took place probably during Jesus' public ministry. How many swords must have pierced her heart! We can only imagine Mary's sorrow and experiences of bitter pain and distress over the humiliating and brutal treatment of her son during His passion. Yet she stood by him, did not try to stop anything, let it happen...

When Jesus fell and was whipped as He carried His cross, Mary must have wanted to take Him in her arms, bind His wounds, give Him comfort, to take on some of His burden herself. Yet she could not. This was her child, but He was the Son of the Father, and he followed Him even as the powers that be sought to destroy him and his threat to their temporal and religious authority. Mary accompanied Him, but could not relieve His suffering directly. She had to suffer with Him, only able to hold Him after His death, as shown in the sorrowful Pieta. Terrance Klein writes: "In history, she holds Jesus, directing attention away from herself and towards her son..." Further, "God made the Virgin silent and vulnerable, like so many women before and after her. Ponder and pray that you be moved by what you see: the silent vulnerability of God made flesh in the womb of the Virgin"[8]. These lines beautifully express how Jesus was rendered increasingly vulnerable to the worldly forces bent on destroying him and his mission, while His mother expressed her vulnerability and sadness in silence. Both suffered, both endured. It is so different in our times, when sometimes people decide to relieve their suffering by asking to have their lives ended, instead of seeking strength from their relationships and consolation from God and from Mary, Mother of God.

---

[8] Klein, Terrance. *America*, December 19, 2019.

III. Mary and Bioethics

Many questions in bioethics arise from our desire to 'fix' something that appears to us as a problem. For example, infertility is a difficult cross to bear, but reproductive technologies now offer ways to allow people the experience of carrying a child and enjoying the fruits of parenthood. Unfortunately, this gift can come at quite a cost, not only in terms of where a couple might place their financial resources, but in regards to the moral aspects of the activities. For example, there are many ethical aspects to consider in a procedure such as in vitro fertilization: the method itself, which separates the procreative and unitive dimensions of marriage, the use of donor gametes, the genetic testing of embryos before transfer to the womb, the discarding of 'flawed' embryos, the freezing of 'spares,' its use outside marriage, the use of surrogates, and so on. All these are involved in fulfilling the desire for a child, at almost any cost. The reproductive technology field is so wide and covers so many issues and services that it is now described as an 'industry.' This is hardly in keeping with human dignity.

Reproductive technologies can acquire for us what we do not have naturally. Mary 'pondered in her heart,' but quietly accepted the Father's will. We do not do this: we move forward to achieve what we want, very often dismissing or not reflecting enough upon serious ethical questions, but creating and terminating human life for our own desires.

Infertile couples could learn from Mary and Joseph here. Mary's Fiat! resulted in her acceptance of a life she did not ask for, something she would not have dreamed possible. Her total surrender to what God asked of her is unparalleled in history, standing as a model of the Father - daughter relationship, in anticipation of the Father - Son surrender. Her commitment to the will of the Father is naturally difficult for us to comprehend, being a deep challenge to our autonomy, which many people today view as unlimited. Joseph's commitment to supporting her and her pregnancy represents another 'let it be!' on his part. Infertility can be an incredible burden to bear, highlighting once more the natural inclination of a man and woman in love to form their own domestic circle, and acceptance of the disappointing reality that this will not happen is incredibly challenging. Looking to Mary and Joseph would help people see that acceptance and surrender to the Will of the Father is necessary, eventually, rather than engaging in a relentless pursuit of what they cannot have. There are other ways of forming family or exercising parental roles through adoption or through the helping professions, and people can be invited to consider these possibilities.

The question of accepting infertility is perhaps even more difficult today when couples see that reproductive technologies have made parenting a possibility for same-sex couples through the use of IVF and donated gametes, and, in the case of a male couple, host parents (surrogates). Catholic teaching is called upon to witness to the dignity of natural conception and the beauty of the inseparability of the procreative and unitive aspects of the spouses' marital acts, while pointing out the indignities suffered by embryos and sometimes by surrogates when pregnancies fail or the fetus is found to have a genetic illness.

If we compare the events in a lowly stable at Bethlehem with technological approaches to reproduction and the ensuing moral problems, it is clear how far society has wandered away from God's plan and from Mary and Joseph's example of obedience and trust, lowliness and acceptance of His will, even in difficult circumstances. Mary and Joseph discovered there was no room at the inn for them, and their example of obedience and trust finds little room in the inns of today. Those of us who pray the Lord's Prayer need to re-intensify Mary's approach to help in these cases: Thy will be done on earth as it is in heaven!

IV. Mary and Gender Ideology

Many theories exist concerning gender identity today, mostly claiming that it is a social construct that does not depend on a person's physical male or female body. In this view, gender is 'assigned' at birth, based on obvious bodily characteristics, whereas the concept of being 'transgender' is based on a person's felt experience that he or she is really a member of the opposite sex, but occupies the wrong body. The given body is perceived as a mistake and is unwanted, and therefore is rejected in favour of altering it to achieve the body of the opposite sex, at least to varying degrees.

Transitioning to the appearance of the opposite sex is made possible at early stages by using puberty blockers to inhibit the development of natural hormones and the changes they bring about, and by the use of drugs to provide secondary sexual characteristics of the opposite sex, such as facial hair or removal of it, voice changes, and so on. Sexual re-assignment surgery is sometimes sought later by those who want to remove their primary sexual characteristics, in order to replace them with surgically-constructed body parts of the opposite sex.

Catholic teaching calls us to honour and accept the body in which we come into the world, seeing gender ideology and practices as completely foreign to the view that men and women are created in God's image as male and female. It is important to note that gender dysphoria is still included as a psychiatric disorder in the American handbook of psychiatric disorders (DSM-5), perhaps since, unfortunately, suicidal ideation very often accompanies the experience[9]. This is a serious condition, and people experiencing such dysphoria deserve good treatment and the best counselling possible.

---

[9] American Psychiatric Association. *Diagnostic and Statistical Manual of Mental Disorders, 5th Edition: DSM-5.* American Psychiatric Publishing, 2013.

Catholic teaching reveres the body as gift, and a gift that must be accepted. Pope John Paul's work in Theology of the Body extols the idea of gift – a gift which we should be grateful to accept and develop, as told in the parable of the talents. Saint Pope John Paul II expressed the moral opinion that radical surgery to remove non-diseased body parts counts as mutilation. Pope Francis notes that some responses to gender dysphoria are ideological, and he reiterates the statement found in the *Catechism of the Catholic Church*: "Everyone should accept his own body"[10].

Rejection of our given body speaks an untruth about God's gift. It speaks an 'Adamic' word in declaring that the owner of the body knows better than God: this is not my body; I dissociate myself from this body and want to start procedures that will make me a member of the opposite sex, to which I feel I belong, present bodily characteristics notwithstanding. The Church is clear that people who feel this way are not to be rejected but helped, but at the same time continues to point out the objective wrong in denying the gift of the body, and reminds people that a body can be altered in many ways to resemble a person of the other sex but can never make the person be the other sex in all it fullness. It has never been part of Catholic tradition teaching that our bodies are ours and ours alone, with which to do whatever we want. Puberty blockers, ongoing hormonal treatment and sex re-assignment surgery to bring about the desired change are ultimately irrational in a distorted attempt to control one's life and fulfill one's desires. Again, this is so far from Mary's response to her situation – the Father asks, and Mary says yes. She did not resist in favour of controlling those aspects of her destiny, she made no attempt to reject the request in order to realize her own desires.

The Church continues to urge us to accept God's gift, to accept the body we are in, because it is essential to our being who we are. One wonders if anyone could be happy or content, feeling in constant opposition to his or

---

[10] *Catechism of the Catholic Church,* Section 2333.

her physical being. Long-term studies yield differing views on this, but more are needed to enable reasonable conclusions.

The gender question is on a completely different plane from other bioethical dilemmas such as contraception, abortion and reproductive technologies, but it shares a common factor in that they have all made motherhood and parenthood incredibly complex. St Pope John Paul II's Theology of the Body stands in stark contrast to views that deny the beauty of each body as God's gift, but which seek to re-create a body according to their own human desires. So, too, Mary's role of obedient acceptance was completely un-self centred: to her, God's request was to be accepted; obedience and trust were her standards, and she cried out later how her soul 'magnified' the Lord. It was never about her and her wishes and desires; everything reflected the Father's will. So many of us have lost this outlook: of course it's about us, and about what we want. Adam and Eve still live!

V. In the light of the Holy Family: Human Love

Men and women have been attracted to each other from the beginning of time: it is evident in the sex drive, the drive to procreate and to satisfy human needs for intimacy, for closeness, caring and stability. Relationships built on these factors are part of the Catholic 'natural law' approach, i.e., it is intrinsic to human nature to manifest these factors, which have consequences and benefits for families and society. The joys of a mutually satisfactory sexual and spiritual life, resting on the foundation of spousal fidelity, love and consideration tend to spill over into parenting, providing a stable and happy childhood for the couple's children. Most people are probably not thinking specifically about parenting when they begin to date and develop some relationships. "Choosing a mate" sounds archaic, but there are many studies that show there is already an undercurrent of that reality in early stages, even if people are not consciously aware of that. Yet our sexuality is so fundamental that it would be strange if we did not realize at some level that we are seeking the 'right one' to share our procreative and unitive hopes and dreams reminiscent of the longing found in the Song of Songs.

Too often, fulfilling our procreative desires, even in marriage, is treated as a decision for us alone to make, and is a matter of yoking technology to our own purposes, whether using contraceptives to prevent pregnancy or IVF to achieve it. It's about what we want: what we want is the most important matter. New life is not seen as God's gift, but about being brought about when we decide. On the other hand, those who see new life as truly God's gift have a different approach. There is an openness and desire to bring a new being into the world, more space in one's heart for one more person to love, more acceptance of fulfilling the needs of a new life, more focus on the little one than on personal needs (at least for a time!), more trust that matters will work out well, more giving of oneself. *Gaudium et Spes* describes parents as cooperating with God in bringing new life into the world and that is a marvellous perspective!:

> Parents should regard as their proper mission the task of transmitting human life and educating those to whom it has been transmitted. They should realize that they are thereby cooperators with the love of God the Creator, and are, so to speak, the interpreters of that love[11].

This approach, looking to God as Creator, serves to shape people into parents, and into better versions of themselves: they will be more selfless, less attracted to consumerism, more aware of the intrinsic value of life, more grateful to God for making us capable. That is not to say that adopting this stance is always straightforward, that parenting is a 'breeze,' that children are never any bother, that sometimes thoughts about falling behind in a career do not occur: these are all too human thoughts, but they most often dissipate in the routine of providing a caring family life, knowing we are being helped

---

[11] *Gaudium et spes*, n. 50

by the grace of God through our openness to the Sacrament we continue to celebrate with each other.

Gaudium et Spes reminds us:

> Authentic married love is caught up into divine love and is governed and enriched by Christ's redeeming power and the saving activity of the Church, so that this love may lead the spouses to God with powerful effect and may aid and strengthen them in the sublime office of being a father or a mother. For this reason Christian spouses have a special sacrament by which they are fortified and receive a kind of consecration in the duties and dignity of their state. By virtue of this sacrament, as spouses fulfil their conjugal and family obligation, they are penetrated with the spirit of Christ, which suffuses their whole lives with faith, hope and charity. Thus they increasingly advance the perfection of their own personalities, as well as their mutual sanctification, and hence contribute jointly to the glory of God[12].

Husband and wife are encouraged to see themselves as domestic church, with a mission from the larger Body to be a light to their neighbours. This is an important part of the Sacrament of marriage. It was never meant to be for the two people alone in isolation, but as a dynamic family relationship that creates its own small church while contributing to the growth of the larger Body of Christ.

This reflects a Trinitarian aspect in marriage as Sacrament, in that the love between the couple brings a third person or more into being, while forming a spiritual bond that both binds and transcends them. *Gaudium et spes* had much to say about this family dynamic:

---

[12] *Ibid.* n. 48.

Thus the Christian family, which springs from marriage as a reflection of the loving covenant uniting Christ with the Church, and as a participation in that covenant, will manifest to all men Christ's living presence in the world, and the genuine nature of the Church. This the family will do by the mutual love of the spouses, by their generous fruitfulness, their solidarity and faithfulness, and by the loving way in which all members of the family assist one another[13].

This is a wonderful affirmation of the role of parents in their mission in the world, and, of course, added to that is the model of the Holy Family. As Mary showed us, do we ponder these things in our hearts? Do we say yes to God, no to self? Do we see the natural beauty of procreation? Do we respect our personal and familial limitations, and, for example, learn to use Natural Family Planning as a way of being responsible parents? What did both Mary and Joseph do? They were both open, obedient, trusting, patient, concerned, non-interfering, not self-seeking, present when needed, and ultimately it would seem that, after Joseph died, Mary suffered with her Son, seeing him beaten, whipped and crucified, abused with taunts and jeers. What mother could bear it? Scripture tells us she did, summarized later in these words: *Stabat mater dolorosa, iuxta crucem lacrimosa, dum pendebat filius* (The grieving Mother stood weeping beside the cross where her Son was hanging[14]).

Mary represented the Church to be founded in His name at the foot of the Cross, gathering him into her arms as she had when he was her child, no doubt murmuring a final earthly farewell, burying Him in the tomb and then rejoicing in his Resurrection. There are so many lessons for us as individuals and as parents through her example! She knew when to speak and when to

---

[13] *Ibid.*

[14] Translation courtesy of the following website: https://www.stabatmater.info/english-translation/.

stay silent; she trusted in God's will when her Son seemed to be a failure in the eyes of the world; during the Passion, she knew her presence was essential yet she stood back and let matters take their course. This can only be done with the maturity born of experience and openness to the Spirit in developing insight and wisdom that allowed her to suffer with Jesus while trusting there would be fulfillment of some kind after all that suffering. Jesus' resurrection from the dead was her (and our) reward and her experiences of trust and fulfilled belief can teach parents that there are experiences that must be experienced fully and that we learn slowly by facing reality and not pursuing distorted personal desires.

Mary's example, together with that of Joseph, gives us a way to follow in married and family life in general, but also gives us a path to guide decisions we may have to make in the life and death matters of bioethics. We do not want to be Adam and Eve, pursuing our own path and ignoring God's Word to us. Mary, in particular, helps us see a better way, a way of saying 'yes!' to God's will, and then trusting that, despite the difficulties sometimes, we are on the path of salvation.

Chapter Six

Mary's Help in Answering Today's Anthropological Questions

General Introduction to Chapter Six: Male and Female. There is a concreteness to the action of God. 'If the concrete were not essential to Christianity, if the Lord were not truly flesh but "merely the product of an idea," then Mary might have been only spiritually pregnant, Adrienne reasons; but the reality of the incarnation requires that she "feel his weight in her body and, after His birth, in her arms"'[1]. Indeed, Hans Urs von Balthasar makes this very clear himself when he says that in the *Incarnation* 'we are of course dealing with a genuine physical fecundation'[2] in which 'the Spirit - faithful to the task received from the triune Father – is primarily an envoy, the obedient bearer of the divine seed and of its actualization in the womb of Mary's Yes'[3]. In other words the word of God is visibly effective in bringing to exist what did not exist.

Just as in the beginning the word of God brought creation to exist so the *Incarnation of the Son of God* shows forth the fruitful action of the Holy Spirit. While, then, this generation may argue that man, male and female, is

---

[1] Michele M. Schumacher, "A Speyrian Theology of the Body", quoting from Adrienne von Speyr, *Maria in der Erlosung*, 35, p. 264 of pp. 255-292 in the book *The Virgin Mary and Theology of the Body*, edited by Donald H. Calloway, Pennsylvania: Ascension Press, 2007.

[2] *Explorations in Theology: V: Man is Created*, translated by Adrian Walker, San Francisco: Ignatius Press, 2014, p. 177.

[3] *Explorations in Theology: V: Man is Created*, p. 179.

a provisional project which may be adapted at will, what is actually astonishing is the unity-in-diversity of precisely those human relationships which originate with the beginning of man's being, male or female. Indeed, just like the triangular based tetrahedron which is a basic building block of the universe, the triangular relationship of God, man and woman is amazingly capable of building up a marvellously complex society from the simplest complementarity, as it were, of man and woman: a complementarity which reaches from the depths of eternity into the foundations of the beginning and is again taken up in the wondrous work of salvation drawing on the virginal manhood of Jesus Christ and the mystery of the spousal mystery of the Virgin Mary. When, according to St. John Paul II, God pondered the divine 'We' - man and woman came forth:

> 'Before creating man, the Creator withdraws as it were into himself, in order to seek the pattern and inspiration in the mystery of his Being, which is already here disclosed as the divine "We". From this mystery the human being comes forth by an act of creation: *"God created man in his own image,* in the image of God he created him; male and female he created them" (*Gen* 1:27)' (*Letter to Families*)[4].

The dialogue of the sexes is, therefore, a part of the very revelation of God: a revelation which, in a sense, involves the unity-in-diversity of human relationships; indeed, that it is necessary to ponder the mystery of man and woman anew: anew because of the multitude of ways that this "Icon" of the Blessed Trinity is being disfigured and needs to be reclaimed to be the healing anointing it is.

---

[4] The English title of this letter is given in the text as the Latin title is less well known, which is *Gratissimam Sane*.

Chapter Six: Introduction: Being-in-Relationship. Mary is pre-eminently the perfect created creature; and, therefore, in her perfection lies the secret of a wholly relational being who is wholly human. Mary's "relationality", however, does not exhaust her individuality but, rather, her relationality is a vehicle of her individuality: her individuality is profoundly expressed in her relationality. Mary is a woman of archetypal relationships - she is both in the midst of the Blessed Trinity and in the midst of mankind.

In the times in which we live, then, we are in an identity crisis involving the whole human race: Is man, male and female, an irrevocably relational being or is each of us a project, as it were, without any ontological roots in the human race? Thus there is a need for a foundational account of human identity: an account that takes up everything that is characteristic of human being and shows forth its coherence: a coherence which begins from conception and unfolds throughout all the relationships of life. In view, then, of the Church's deepening reflection on the reality of the woman there is an implicit providentiality to it: a timeliness whereby the questions that have arisen in our times may well find that there are unexplored answers in the mystery of Mary.

In particular, then, Mary is conceived through the spousal love of Anne and Joachim and is, therefore, conceived within the Jewish tradition that her destiny will transcend. Mary's own marriage to Joseph, transformed as it is by the coming of Christ, both bridges the relationship between Judaism and Christianity and universalizes, as it were, the new reality of the sacrament of marriage revealing an interpersonal nature to the action of God in the history of the covenant which it transfigures. The intersection of the old and the new dispensation is, in a way, pre-eminently expressed in Mary being conceived without original sin – a sin transmitted in the flesh deprived of original grace – and, in herself becoming the Mother of God, comes to be in the very epicentre of the generation of the "renewal of man": the complementary centre to the coming of Christ.

The mystery of Mary, then, offers a new beginning to understanding the mystery of man, male and female, from the moment of conception's inclusion of each one of us in the history of salvation. Thus the following sections are: Man, male and female, Christ and Mary (I); Mary and Bioethics (II); Mary and Gender Ideology (III).

Man, male and female, Christ and Mary (I)

In one sense this may seem a statement of the obvious or even a lopsided title, giving undue emphasis to Mary in that she is female and yet singled out for particular attention along with Christ; however, the point is that not only is there an expanse of time between Eve and Mary but there is, too, the constancy of the identity of woman as indeed there is in the masculinity of Adam and the coming of Christ: 'the final Adam' (*Gaudium et Spes*, 22). In other words, in the prophetic anticipation of the 'new Eve' (*Lumen Gentium*, 63) Mary is an expression of the manifest will of God that 'woman' is a perennial gift of human nature and not a passing or ephemeral creature; and, similarly, the coming of the Son of God in the flesh of a true man signifies the enduring nature of manhood. If, to paraphrase St. Thomas Aquinas, grace builds on nature, grace does not re-invent nature; and, therefore, there is implicit in the creation of the whole human being a logic of relationship: Mary is the daughter of Joachim and Anna; Christ is the son of God and the son of Mary, whose foster father is Joseph.

On the one hand Mary and Jesus Christ point back to that prophetic anticipation of their coming when God promised a decisive response to the serpent despoiling the original gift of man and woman and their relationship to one another and to creation; indeed, God says: 'I will put enmity between you and the woman, and between your seed and her seed; he shall bruise your head, and you shall bruise his heel' (Gn 3: 15; cf. also *Lumen Gentium*, 55). "O Woman" (Jn. 2: 4), let us recall, is the expression that Christ uses when addressing His mother at the marriage feast of Cana in response to the possibility that He will give a sign of who He is while, at the same time, helping a newly married couple with a wine better than the one they had (cf. Jn. 2: 10). On the other hand, 'a woman clothed with the sun' (Rev. 12: 1) who 'brought forth a male child, one who is to rule all the nations with a rod of iron' (Rev. 12: 5) communicates a sense of the perennial nature and value of the particularity of each person being male or female (cf. Gn. 1: 27); indeed, in the words of Christ Himself: "Have you not read that he who made them from the beginning made them male and female ..." (Mt. 19: 4).

Not ignoring, then, the literal sense of God creating a man, Adam, and a woman, Eve, and the possibility of their union being a symbol of the union between God and man, there is even a sense in which the symbolic significance of 'man' and 'woman' is taken up into a new, literal-spiritual sense, in that the Son of God became man in Jesus Christ, the 'bridegroom' (Mk. 2: 19 and 20), whose bride is the Church. In other words, whatever the storms and swirls of contemporary thinking, feeling and choosing, there is indeed a perennial nature in the very creation of 'male and female' human beings. If, then, it is true, that 'male and female' are authoritatively normative, then it follows that there is a perennial significance to the "relationship" between them; and, indeed, while the history of salvation is an expression of the enduring significance of that relationship there remains the significance, as it were, of "relationship itself".

In view, however, of the widespread destruction of human life it could be argued that we are living in a relational winter in which, for a variety of reasons, the reality of relationship is denied. One of the tragically recurring ways in which this "relational winter" is expressed is in the denial of parenthood: in the denial of parenthood as a relationship between father, mother and child: in the denial of the human relationships implied in deliberate abortion. Recovering a sense of relationship, however, includes recovering a sense of spousal and parental love and, indeed, recovering a sense of being a loved child of God. More deeply, however, it means pondering the mystery of relationships in the heart of our Creator-God: the mystery of the Blessed Trinity.

In a word, then, the existence of man and woman is a constituent expression of the nature of the human race: a race of relational human beings; and, if confirmation were needed, there is the eco-system in which we live and the relational nature of all reality expressed, as it were, in the structure of the universe. Maybe the denial of God is at the root of a relational confusion in the contemporary world; indeed, as it says in *Gaudium et Spes*, 'once God is forgotten, the creature is lost sight of as well' (36).

Why do Bioethical Questions need the Help of Pondering the Mystery of Mary? (II)

Cardinal Ratzinger, as we shall see, uses a terminology that starts to help us identity the terms of the problem as well as the possibility of a solution.

Pondering the mystery of Mary enables us to behold the whole woman as a word of God (IIi)

On the one hand 'If, therefore, Christ and *ecclesia* are the hermeneutical center of the scriptural narration of the history of God's saving dealings with man, then and only then is the place fixed where Mary's motherhood becomes theologically significant as the personal concretization of the Church'[5]. 'But this biological fact [of Mary's motherhood] is a theological reality, because it realizes the deepest spiritual content of the Covenant that God intended to make with Israel'[6]. On the other hand "The 'biological' and

---

[5] Cardinal Ratzinger, "Thoughts on the Place of Marian Doctrine and Piety in Faith and Theology as a Whole", *Communio*, 30 (Spring 2003): p. 155: https://www.communio-icr.com/files/ratzinger30-1.pdf.

[6] Ratzinger, "Thoughts on the Place of Marian Doctrine and Piety in Faith and Theology as a Whole", p. 155.

the human are inseparable in the figure of Mary, just as are the human and theological"'[7].

In the first place Mary is understood in terms of her motherhood: in terms of how her relationship to her Son roots her in the salvific relationship between 'Christ and [His] *ecclesia*'. There seems to be both an embracing of 'Mary's motherhood' and a development of it in terms of it being 'theologically significant as the personal concretization of the Church'. In terms, however, of when the author says that this 'biological fact [of Mary's motherhood] is a theological reality', there seems to be an almost too abrupt transition from 'biological fact' to 'theological reality'. In other words there is no mention of the psychological reality expressed in the 'biological fact' of Mary's motherhood. Conversely, it could be argued, stating the bare fact of the 'biological fact [of Mary's motherhood] could be a way of radicalizing our perception so that we realize the wholesomely encompassing incorporation of Mary's motherhood; indeed, that her motherhood has taken on a kind of universalizability in virtue of who her son is. At the same time, however, 'motherhood' is not a bare 'biological fact'; it is, in its reality, a biologically inscribed psychological reality[8]. In other words, just as there is a spousal relationship in the procreation of a child there is a relationship of motherhood on the child coming to exist as indeed there is a relationship of fatherhood; and, therefore, even in the case of the *Incarnation* there is the reality of Mary's relationship to God which precedes the reality of the *Incarnation*. The *Incarnation* does not replace Mary's relationship to God: it expresses that relationship; and, in virtue of the religious nature of her marriage to Joseph, the validity of that marriage remains. The coming of Christ is an expression of Mary's relationship to God which, as it were,

---

[7] Ratzinger, "Thoughts on the Place of Marian Doctrine and Piety in Faith and Theology as a Whole", p. 157.

[8] Cf. Francis Etheredge, *The Human Person: A Bioethical Word*, St. Louis: En Route Books and Media, 2017, e.g. p. 39: http://enroutebooksandmedia.com/bioethicalword/.

unfolds in her relationship to the Son of God, to St. Joseph and indeed to all of us (cf. Jn. 19: 26-27).

At the same time, the psychological structure of human being, spousal love and motherhood are implied in the following expression: "The 'biological' and the human are inseparable in the figure of Mary, just as are the human and the theological". Indeed, in the words of Mary, on finding the child Jesus after three days of searching with Joseph, there is a simple articulation of the wholesome reality of these relationships: "Son, why have you treated us so? Behold, your father and I have been looking for you anxiously" (Lk. 2: 48). To which Jesus' reply adverts to His "hidden" identity: "Did you not know that I must be in my Father's house?" (Lk. 2: 49). But the Gospel goes on to say: 'And he went down with them and came to Nazareth, and was obedient to them' (Lk. 2: 51). Thus God took up His own "creation" of the motherhood of Mary and the foster fatherhood of St. Joseph showing, through His own acceptance of Joseph as a father and Mary as a mother, the humanization of His humanity that comes through relationships to parents.

Almost, then, as a counterpoint to the expression that the 'biological fact [of Mary's motherhood] is a theological reality' is the further elucidation that the "'biological' and the 'human' are 'inseparable in the figure of Mary". But the 'biological' and the human are not 'inseparable' in virtue of it being Mary, the Mother of God, except in so far as this is the perfect realization of human womanhood; rather, the 'biological' and the 'human' are 'inseparable' in virtue of the nature of human being: the ontological whole of woman entailing the biologically rooted psychological structure of womanhood. What is particularly clear in Mary, however, in that the 'human and the theological' are inseparable is precisely the sense that the wholeness of human being entails a theological word: a word of God.

Exploring the "theological" and the "psychological" in the mystery of Mary (IIii)

On the one hand, 'Perhaps more than men, women *acknowledge the person,* because they see persons with their hearts' (St. John Paul II, *Letter to Women*, 12).

On the other hand, 'this biological fact [of Mary's motherhood] is a theological reality, because it realizes the deepest spiritual content of the Covenant that God intended to make with Israel'[9].

Taking, then, the words and sign of the first explicit Covenant it says in *The Book of Genesis*: "Behold , I establish my covenant with you and your descendants after you, and with every living creature .... that never again shall all flesh be cut off by the waters of a flood and never again shall there be a flood to destroy the earth .... [and] I set my bow in the cloud, and it shall be a sign of the covenant between me and the earth" (Gn. 9: 9-11, 13).

In the first place covenant means relationship, promise and outward sign: 'I establish my covenant with you and your descendants after you' (relationship); 'and never again shall there be a flood to destroy the earth' (promise); and, finally, 'I set my bow in the cloud, and it shall be a sign of the covenant between me and the earth' (sign). It is clear, then, that God establishes the covenant with a universal embrace of both the 'earth' and with 'you and your descendants'; and, indeed, it is like a binding universal embrace in that the covenant is between God and the earth and 'you and your descendants'.

If Mary's motherhood, then, 'realizes the deepest spiritual content of the Covenant that God intended to make with Israel' then God has "gone to the limit", as it were, in the *Incarnation* of the Son of God in the womb of Mary; and, indeed, in the response of Mary's faith that believed 'with God nothing

---

[9] Ratzinger, "Thoughts on the Place of Marian Doctrine and Piety in Faith and Theology as a Whole", p. 155.

will be impossible' (Lk. 1: 37) there is, literally, the embodiment of the human-divine relationship in taking the flesh of man from Mary. In other words, in the very nature of bearing a child the *Incarnation* of the Son of God involved a relationship to Mary that is both specific to her and universal: 'For, by his incarnation, he, the son of God, has in a certain way united himself with each man' (*Gaudium et Spes*, 22). Maternity has, as it were, expressed and entered eternity. In other words, the whole principle of relationship is expressed pre-eminently in the Son of God being the son of Mary; and, therefore, the covenant relationship is now expressed in terms of the "flesh of Christ": the flesh of Christ now expresses, from the *Incarnation* onwards, the indestructible nature of redemption in terms of the union of the two natures, of man and God, in the one person of Jesus Christ (cf. CCC, 464-469). Thus what Christ undergoes, He undergoes in a way that transfigures the relationship between man and God entered into in the *Incarnation*; and, therefore, in that the *Incarnation* is not an abstract event but the profoundly human relationship of mother and son, then the transformation of that relationship is a sign of the transformation of all relationships in Christ.

If 'women *acknowledge the person,* because they see persons with their hearts' (St. John Paul II, *Letter to Women*, 12) is based on an ontological reality perhaps the ontological reality it is based on is the fact that the woman is, as it were, the "place" of relationships; indeed, just as the whole event of the *Incarnation* introduced Mary "in" and "to" the Blessed Trinity, so in her virginally betrothed womanhood becoming motherhood, Mary becomes the "place" of the meeting of all human relationships and God.

Exploring the "biological" and the "psychological" in the "woman" Mary (IIiii)

The nature of maternity, then, involves the very flesh of the woman as the "receptacle" of life, literally in the fertilized ovum being the child she bears.

Conception begets, too, the beginning of the biologically inscribed psychological relationship of mother and child, which already begins to be expressed in Mary's expectation of the possibility of her immanent motherhood (cf. Lk. 1: 31-38). Ordinarily, the psychological reality of fatherhood is inscribed in the transmission of life being fruitfully received in the begetting of the child; however, in the nature of Mary's virginal conception of Christ, the presence of God is actively present in bringing about the mystery of the *Incarnation* through the "overshadowing" of 'The Holy Spirit' (Lk. 1: 35). In other words, the "Fatherhood of God" is mysteriously expressed in the *Incarnation* in a way which brings about the ensouled conception of Christ in the moment of the transcendental act of the *Incarnation* of the Son of God; however, the spiritual nature of the Fatherhood of God while bringing about the reality of the *Incarnation* does not replace but facilitates, as it were, the foster-fatherhood of Joseph (cf. Mt. 1: 20-21). As it says in St. Luke, after Christ has referred to being "in my Father's house" (Lk. 2: 49) and therefore acknowledging a fatherhood different to that of Joseph's, he submits to their parenthood: 'And he went down with them and came to Nazareth, and was obedient to them; and his mother kept all these things in her heart' (Lk. 2: 51).

Mary, then, expresses both her own motherhood and Joseph's fatherhood when she says: 'your father and I have been looking for you anxiously' (Lk. 2: 48). Mary's motherhood is "spousally" ordered to Joseph's fatherhood such that Jesus Christ experiences the dynamic parental relationships of mother and father; and, therefore, Mary's motherhood is not an "excluding" kind of motherhood which rejects Joseph's spousal and parental relationships. In other words, even in view of the extraordinary nature of the *Incarnation* and the reality of Mary's motherhood, Mary is open to the graced relationship to Joseph and his fatherhood: graced because of it being the will of God (cf. Mt. 1: 20-21). At the same time Mary entered into the relationships to do with her family and people, as when she visits (Lk. 1: 40) her 'kinswoman Elizabeth' (Lk. 1: 36) and cared for other children to the point that they were regarded as the 'brothers and sisters' of Jesus (Mk. 6: 3); and, moreover, Mary readily recognizes her own and Joseph's psychological reality when she speaks of them searching 'for ... [Jesus] anxiously' (Lk. 2: 48). Furthermore, it is Mary who perceives the need for wine at the wedding feast at Cana, observing discreetly that they have run out and addressing Jesus: "They have no wine" (Jn. 2: 3). In other words, although sketched with few strokes, Mary's relationship to Jesus Christ is rooted in the reality of her "biological" and "psychological" relationship to others; and, at the same time, she expresses a wholesome recognition of herself in the midst of needs which transcend her own ability to help and yet lead to drawing on her relationship to her son.

A preliminary answer, then, to the possibility of a Marian "bioethical word", as it were, is that Mary's vocational fulfilment of being woman, spouse and mother takes up all that is profoundly human. Indeed, in a certain way, the fruitfulness of Mary's motherhood is a kind of sign to Joseph and to men in general that there is a poverty about masculinity which needs the answering power of God; and, therefore, the words of Eve possess a kind of perennial truth that helps to express the humility of being human: 'I have gotten a man with the help of the Lord' (Gn. 4: 1). It is possible that this humility is essential to a real and profound appreciation of the reality of human being. In other words, without explanation, it may well be that the "good" of creation (cf. Gn. 1: 4, etc. and 31) entails a normative awareness of male and female complementarity, the transmission of life through spousal love, the welcoming receptivity and nurturing love of the mother, the paternal love of the father and the dynamic of marital love informing the interactive love of the parents. The very fact that the *Incarnation* of the Son of God involved becoming a true human embryo is recognition that there is a "normative value" in the very conception of Christ taking up the whole process of embryonic maturation and a relationship to both Mary and to her husband, Joseph,

The real challenge, then, to recognizing human rights is the recognition of the human nature that unfolds coherently through all the relationships of life; and, therefore, there cannot be a coherent account of human rights if there is no coherent account of the human being-in-relationship.

Mary and Gender Ideology (III)

The following excerpts have been chosen by way of taking what appears to be the deepest problem expressed in terms of gender ideology: the problem of being 'free from every predetermination linked to their essential constitution'.

'In order to avoid the domination of one sex or the other, their differences tend to be denied, viewed as mere effects of historical and cultural conditioning. In this perspective, physical difference, termed *sex*, is minimized, while the purely cultural element, termed *gender,* is emphasized to the maximum and held to be primary'[10]. 'Further, the concept of gender is seen as dependent upon the subjective mindset of each person, who can choose a gender not corresponding to his or her biological sex, and therefore with the way others see that person (*transgenderism*)'[11].

---

[10] Excerpt from the "Letter to the Bishops of the Catholic Church on the Collaboration of Men and Women in the Church and in the World", 2: http://www.vatican.va/roman_curia/congregations/cfaith/documents/rc_con_cfaith_doc_20040731_collaboration_en.html.

[11] Excerpt from the document "Male and Female He Created Them": Towards a Path of Dialogue on the Question of Gender Theory in Education, by the Congregation for Catholic Education, Vatican City, 2019; text from the article "New Vatican Document Provides Schools with Guidance on Gender Issues", June 10th, 2019: https://zenit.org/articles/new-vatican-document-provides-schools-with-guidance-on-gender-issues/, 11.

Arising, then, out of the context of a legitimate reflection on 'women's roles' is 'the human attempt to be freed from one's biological conditioning'; and, therefore, it is argued that 'human nature in itself does not possess characteristics in an absolute manner: all persons can and ought to constitute themselves as they like, since they are free from every predetermination linked to their essential constitution'[12].

By contrast, 'The primary, foundational contact between mother and child, indeed, the symbiosis between them, is by no means purely "biological", especially given contemporary reflection on "being with". The essential event is a spiritual one'[13].

Mary and Gender Ideology (IIIi): "Being with"

There are, then, a number of points to discuss beginning with the relationship between *sex* and *gender*; and, accordingly, let us take the following terms: *sex* is defined as 'physical difference' while *gender* is the primary, 'purely cultural element'.

On the one hand *sex* is a physical difference; and, therefore, there is a question: What is the nature of this 'physical difference'? Being a woman, as indeed being a man, entails a 'physical difference' which is, as it were, a psychological difference; indeed, it is precisely in the contrast between a 'physical difference' and a 'purely cultural element' that the "theoretical chasm" exists. In other words, just as 'physical difference' expresses psychological characteristics so the 'cultural element' proceeds from the nature of human psychology. In other words it is only because of an inadequate anthropology that it is possible to polarize 'sex' and 'gender'.

---

[12] Adapted excerpts from the "Letter to the Bishops of the Catholic Church on the Collaboration of Men and Women in the Church and in the World", 3.

[13] Hans Urs von Balthasar, "Mary in the Church's Doctrine and Devotion", p. 104 of Part II of the book, *Mary: The Church at the Source*, 2005.

In modern terms, the "being with" between mother and child arises out of the "biological" expression of motherhood; and, indeed, "being with" expresses the relational reality which begins with conception in that the mother is precisely the person who is "being with" the child. In the very nature of motherhood, then, there is a recognition of the reality of the person of the unborn child and a relationship of accompaniment; indeed, in the language of the Church today there is a definite recognition of the Church "being with" each person in the course of their Christian life[14]. Therefore, while it is true that there is relationship between Mary and the Church, such that what is true of the one is also true of the other[15], it is also that in the perfection of Mary there is, as it were, not only a translucent 'Ark of the Covenant' in which God dwells[16] but there is, too, a clearer perception of the significance of being 'woman' (Jn. 2: 4 and Jn. 19: 26). In reality, then, there is a kind of running significance between the psychological capacity of "being with" which exists in the very nature of maternity being a "nurturing place" for the dwelling of another, prior to the externalization of this relationship through birth and the nature of the divine maternity of the Church "being with us" in the course of our earthly pilgrimage in the hope of "being with us" in eternal life.

On the other hand, if it is true that the roots of a maternal psychology are a profoundly personal perception of the whole human being, then this capacity to appreciate the whole is itself rooted in an almost synthetic grasp of reality; indeed, while in the context of explaining the nature of Mary's

---

[14] Cf. Francis Etheredge, *The Family on Pilgrimage: God Leads Through Dead Ends*, St. Louis: En Route Books and Media, 2018, pp. 160-163: http://enroutebooksandmedia.com/familyonpilgrimage/.

[15] Ratzinger, "Hail, Full of Grace", p. 66 of *Mary: The Church at the Source*, 2005: 'Mary is identified with daughter Zion, with the bridal people of God. Everything said about the *ecclesia* in the Bible is true of her, and vice versa: the Church learns concretely what she is and is meant to be by looking at Mary.'

[16] *Ibid*, p. 65.

pondering, Ratzinger says: Mary 'translates the events into words and penetrates them, bringing them into her "heart" – into that interior dimension of understanding where sense and spirit, reason and feeling, interior and exterior perception interpenetrate circumincessively'[17]. This is exactly the kind of contemplative wholeness which was expressed by the mother of seven martyrs, who was also martyred, when she said: "It was not I who gave you life and breath, nor I who set in order the elements within each of you. Therefore the Creator of the world, who shaped the beginning of man and devised the origin of all things, will in his mercy give life and breath back to you again, since you now forget yourselves for the sake of his laws" (2 Mac. 7: 22-23).

By contrast, then, the very witness of the nature of woman to the integrity of human being is going to "open up", as it were, a proportionate pain if the woman does violence to "being with" her child; however, to the extent that she discovers the "being with" of the maternity of the Church, there is the possibility of the healing of her pain[18].

Mary and Gender Ideology (IIIii): Culture and Philosophy

It is argued, however, that 'human nature in itself does not possess characteristics in an absolute manner: all persons can and ought to constitute themselves as they like, since they are free from every predetermination linked to their essential constitution'[19].

On the one hand it seems perfectly reasonable that there is equality in terms of the basic characteristics of human development, whether that be

---

[17] Ratzinger, "Hail, Full of Grace", p. 71 of *Mary: The Church at the Source*, 2005.

[18] Cf. Ratzinger, "Hail, Full of Grace", p. 77 of *Mary: The Church at the Source*, 2005.

[19] Adapted excerpts from the "Letter to the Bishops of the Catholic Church on the Collaboration of Men and Women in the Church and in the World", 3.

education or employment; and, indeed, there is a just desire for the full expression and remuneration of a woman's talents, training and experience. In the event, then, of an almost persecutory sense of prohibiting or frustrating the development of the woman's identity then it would be that there is an injustice to be recognized, challenged and indeed overturned. At the same time, however, this does not itself prejudice the complementary character of the masculine-feminine expression of the whole human race; indeed, if anything, it expresses the possibility of a more fruitful dynamism precisely because of the increasingly "present" reality of the woman.

On the other hand the sense that arises out of a desire to renounce any 'predetermination linked to … [a person's] essential constitution' is that of a kind of destruction of the "givenness" of being male or female; and, therefore, therein lies the problem of the possibility of pain: that the pain of being expressed in terms of being a man or a woman is a psychological pain that is, as it were, turned on the self. This is not, however, a discussion of those expressions of human personhood which possess, as it were, an organic ambiguity and which require extremely careful assessment of the whole person; for, just as the biological expression of human personhood has a psychological interiority, so the possibility of psychological confusion arising out of an organic ambiguity needs addressing[20]. What is at issue here,

---

[20] Cf. "Male and Female He Created Them": Towards a Path of Dialogue on the Question of Gender Theory in Education, 24: In 'cases where a person's sex is not clearly defined, it is medical professionals who can make a therapeutic intervention. In such situations, parents cannot make an arbitrary choice on the issue, let alone society. Instead, medical science should act with purely therapeutic ends, and intervene in the least invasive fashion, on the basis of objective parameters and with a view to establishing the person's constitutive identity.' Cf. The real difficulties of this problem, both within the family and socially, in "Lovingly, a family raises an intersex child – again", by Lindsay Whitehurst, *The Associated Press*, June 9th, 2019: https://toronto.citynews.ca/2019/06/09/lovingly-a-family-raises-an-intersex-child-again-2/.

however, is the very "charged" atmosphere of a cultural climate in which gender confusion is almost induced due to the quasi-philosophical claim that there is no 'predetermination linked to ... [a person's] essential constitution'. On the one hand, then, the adult psychological rejection of male or female identities engendering a climate of confusion in which children develop a kind of "imitative" questioning of their identities is a problem in its own right. On the other hand this climate of confusion is nevertheless rooted in the philosophical claim that there is no 'predetermination linked to ... [a person's] essential constitution'.

Mary and Gender Ideology (IIIiii): A Philosophical Claim

What is it about the claim that the men and women are social inventions, like photoshop products, and that what surrounds us is not real? Is it possible to ask where this desire to label ourselves other than men and women came from? Why do we think that the psychological is not inscribed in the biological framework, as it were, in the development of human being? What is so plausible about the idea that everything can change and be changed? What makes it so easy for us to think that the only reality is what is imagined, invented or changed?[21]

It is necessary, then, to address the philosophical claim that there is no 'predetermination linked to ... [a person's] essential constitution'. If there is no 'predetermination linked to ... [a person's] essential constitution' then what is a person's 'essential constitution'? If being a man or a woman is not a part of a person's essential constitution then why does nature, as it were, transmit human identity in terms of one or the other? Not everyone is called to marriage, to raise children or to the religious life; but, nevertheless,

---

[21] Cf. Christopher Dummitt, "'I Basically Just Made it Up': Confessions of a Social Constructionist", September 17th, 2019, *Quillette*: https://quillette.com/2019/09/17/i-basically-just-made-it-up-confessions-of-a-social-constructionist/.

everyone is both inseparably relational and called to be-in-relationship to others. If, then, the manufacturing ethos is internalized, such that what I am is raw material for what I want to be, then it follows that I look upon myself in terms of being made up of parts and not in fact being a whole. If, therefore, I am not a whole but an assemblage of parts, what new "whole" can I be if all I am is an assemblage of parts? In other words, if I want to be different to what exists – then what will determine the "image" of the whole that I want to be? To think in terms of a radical plasticity to human being is to imagine that biological differences are not an integral part of the person.

A thought experiment: Carnivorous head and vegetarian body or vegetarian head and carnivorous body

On the view that biological differences are external to personhood it is mechanically possible, as it were, to change the appearance of a creature. Changing the outward signs of femininity or masculinity does not make a person masculine or feminine; indeed, it is like claiming that while red traffic lights mean "stop" for the rest of the world, red traffic lights for me, mean "go" – a change in significance that is both claimed individually and advanced as a public change in the sign language system[22]. In other words, irrespective of the social damage to common communication systems, it is claimed that there is no existence prior to naming: that naming identifies what exists without reference to the characteristics of a being-in-existence; and, therefore, it is claimed that I am what I want to call myself to be. Thus, whether or not there is a sign system that says a red traffic signal means 'stop' and a green one 'go', it is claimed that there can be personal alterations of this system that are registered and valid for me and for me only.

To use another analogy it is like grafting the head of a carnivore onto a grazing animal's body. In other words the head does not change the stomach

---

[22] Cf. Etheredge, *The Human Person: A Bioethical Word*, p. 257.

nor the stomach change the head; indeed, just as the head of a grazing animal goes with its stomach so does the mentality of the animal go with how it eats. Thus the mentality of a grazing animal is not going to help a carnivorous animal find food nor, conversely, will the mentality of a carnivorous animal help the grazing animal to eat, little by little, over a long period of time. Over time, the physiological structure of the vegetarian head will atrophy the carnivorous bodily structure that it is not capable of maintaining; as, indeed, the physiological structure of the carnivorous head will atrophy the vegetarian bodily structure that it is not capable of maintaining. What transpires, then, is a kind of discrepancy between the physiological "mentality" of the bodily structure and the differently embedded brain structure of the newly grafted head. Consider, then, adapting other kinds of creatures and, in the process, begin to recognize, conversely, the coherence of bodily structure and embedded nervous and hormonal influences on the animal's habitual mentality. In other words, on the basis of this thought experiment it is possible to imagine the myriad implications of a radical change to "parts" of an animal: Would a nest making bird manage to dig a hole in the ground or would a ground digging animal manage to build a nest in a tree? Thus, by extension, it is possible to recognize that the original animal is a coherent whole: physiology expresses a mentality which is expressed in the natural framework of a creature's life.

The witness of sport

In the words of one young female athlete: "It's not like we're saying that we don't like transgender people," she added. "It's just an equality issue where these girls are trying their absolute hardest to try and get those good things on their college resumes, and then it just gets completely taken away from

Chapter Six 213

them because there's a biological male racing against them"²³.

At the level of international sport the same problem is arising. Although not all the details have been made public, an athlete competing in a woman's race was told that owing to a high level of testosterone there was no compatibility with other women athletes: 'The IAAF has ruled Semenya's testosterone will be too high for the athlete to compete in the 800m events going forward. However, <u>the court ruled</u> that "such discrimination is a necessary, reasonable and proportionate means of achieving the IAAF's aim of preserving the integrity of female athletics," according to the <u>Washington Post</u>'²⁴.

In another ruling, a sport's committee said: '"Our rules, and the basis of separating genders for competition, are based on physiological classification rather than identification. On the basis of all information presented to the Board of Directors for this particular case, the conclusion made, is that the correct physiological classification is male," Paul Bossi, the president of the federation, said'²⁵. At the same time, however, as there is the problem of unfair advantages and difficulties for referees, there is a real risk of injury due

---

[23] May 8th, 2019, The Daily Signal "8th Place, A High School Girl's Life after Transgender Students Join Her Sport": https://www.dailysignal.com/2019/05/06/8th-place-high-school-girls-speak-out-on-getting-beat-by-biological-boys/.

[24] "Semenya's testosterone levels deemed too high to compete in women's 800m", by Nic Zumaran, 12th May, 2019: https://www.bioedge.org/bioethics/semenyas-testosterone-levels-deemed-too-high-to-compete-in-womens-800m/13052?inf_contact_key=5893909908014a5ad3cdbd5af6fb64d67e470d92b8b75168d98a0b8cac0e9c09.

[25] "Olympians' fury after transgender weightlifter's perfect performance", by Staff Writer, News, Saturday, 11th May: https://7news.com.au/sport/olympians-fury-after-transgender-weightlifters-perfect-performance-c-106620?inf_contact_key=8aa00c3cc3ad5403b49c4bc8a41a034fcc0558ed5d4c28cbfab114022b1ec50d.

to sex-based differences: 'Some refs are <u>saying</u> they afraid of being sued as more men claiming to be women join the women's leagues and end up hurting natural-born female contestants with their strength and speed, the *Sunday Times* reported'[26].

In the experience of three people: the emergence of evidence and its questions

Contrasting carnivorous and herbivorous animals does not do justice, however, to the intense suffering which people experience over their sexual identity and, by implication, it is an experience which their families and those around them share in too – both positively and negatively. There is a need, then, for a 'methodology' to guide 'both individuals and communities: to listen, to reason and to propose. In fact, listening carefully to the needs of the other, combined with an understanding of the true diversity of conditions, can lead to a shared set of rational elements in an argument'[27]. Thus the following experiences are recounted and briefly reviewed; however, they are in the context of the broader dialogue on the identity of the human being which has been drawing on the work of Shulamith Firestone and others.

One man was so traumatised by what happened to him both as a boy and as a man that he wanted to become a woman and then went on to identify himself as 'non-binary', as if he was neither male nor female and, finally, he rediscovered the fact that he was a man. What he found so difficult to avail

---

[26] "Women's Rugby Referees Fear Transgender Athletes Will Cause Major Injuries", by Warner Todd Huston, 30th September, 2019:
https://www.breitbart.com/sports/2019/09/30/womens-rugby-referees-fear-transgender-athletes-will-cause-major-injuries/?inf_contact_key=0b125148c4c75bc4c014e07a574eade21b0a3f0fd3ee5d9b43fb34c6613498d7.

[27] "Male and Female He Created Them": Towards a Path of Dialogue on the Question of Gender Theory in Education, 5.

Chapter Six

himself of was the very therapy he now seeks to address his sufferings: 'As a child, I was sexually abused by a male relative. My parents severely beat me. At this point, I've been exposed to so much violence and had so many close calls that I don't know how to explain why I'm still alive. Nor do I know how to mentally process some of the things I've seen and experienced'[28]. In other words there are truly painful experiences that disorientate a person and raise profoundly intricate, delicate and difficult to address identity issues; however, what seems to be at work in the long and often maze-like reality-walk between what disturbs a human being and what helps to bring peace is a reconciling self-acceptance[29]. James said: 'In January 2019 ... I reclaimed my male birth sex'[30].

In the reality of a woman taking testosterone, a hormone that generally stimulates facial and skin hair, along with surgery to remove her breasts, a woman takes on the appearance of a man; however, when considering the possibility of removing her womb, she decided against it because she 'considered a hysterectomy, but never went through with it – partly because ... [she] had not ruled out the possibility of having children'[31]. In other words, as far as this woman went to change her appearance into that of a

---

[28] Jamie Shupe, "I was America's first 'nonbinary' person. It was all a sham":https://www.lifesitenews.com/opinion/i-was-americas-first-nonbinary-person.-it-was-all-a-sham.

[29] Cf. "Ex-transgenders tell US Supreme Court: It's 'abuse' to affirm gender confusion", by Doug Mainwaring, September 27th, 2019:
https://www.lifesitenews.com/news/ex-transgenders-tell-us-supreme-court-its-abuse-to-affirm-gender-confusion?utm_source=LifeSiteNews.com&utm_campaign=d7effc8d70-ProFam_10_01_2019&utm_medium=email&utm_term=0_12387f0e3e-d7effc8d70-402645257&mc_cid=d7effc8d70&mc_eid=b03be40d41

[30] Shupe, "I was America's first 'nonbinary' person".

[31] "Man and boy", by Simon Hattenstone, p. 15 of Weekend Supplement of the year (The Guardian), 20-04-19.

man, she could not bring herself to remove, irretrievably, the possibility of being a mother. In the event of now having had a child, it remains to be seen if, over time, the reality of being a mother affects the identity confusion of the woman and helps her to recognize the goodness of her womanhood; indeed, already, owing to the removal of testosterone from her body, she 'starts having periods again … facial hair gets wispier, … hips broaden, … tummy softens' and she starts 'to speak less from … [the] chest and more from … [the] throat'[32].

In the view of a man taking 'oestrogen' he talks about his body 'still producing testosterone, so the female hormones are having to fight their way past the male ones'[33]; and, in the gloss on the article at the end the commentator says: the '*Author … still lives as a man, but has begun the male-to-female gender transition that will eventually result in becoming a woman*'[34]. It is necessary to take account, then, of Shupe's comment that 'many therapists don't question anything anymore when a patient says he is gender-confused'[35]; and, in view of a media tendency to transmit an unquestioning acceptance of these ideas, there is a need to evaluate what is going on: to think

---

[32] Hattenstone, "Man and boy", p. 15.

[33] David Thomas, "The Wrong trousers: My transgender diary", The Telegraph Magazine, 27th April, 2019, p. 11.

[34] Thomas "The wrong trousers", p. 11: additional comment at the end of the article.

[35] Lisa Bourne, "'The lie was crushing me': First gov't-recognized 'non-binary person' re-embraces his male sex":

https://www.lifesitenews.com/news/james-shupe-first-government-recognized-non-binary-person-re-embraces-his-male-sex?utm_source=LifeSiteNews.com&utm_campaign=7362ef5772-ProFam_4_18_2019&utm_medium=email&utm_term=0_12387f0e3e-7362ef5772-402645257&mc_cid=7362ef5772&mc_eid=b03be40d41.

through the question of identity[36].

There is a difference, however, between altering the appearance of a person and a person becoming other than he or she is in terms of being male or female. The idea, then, that a hormone can determine an adult 'male-to-female' change presupposes that there is no psychosomatic wholeness to the existence of being a man or a woman; indeed, that 'person' and 'sex' are somehow different. In other words, it is possible to be a man that thinks it would be better to be a woman or to be a woman who thinks it would be better to be a man; however, thinking these thoughts does not change the reality that it is a man that thinks of becoming a woman and a woman that thinks of becoming a man. Moreover, the question arises, what is it about becoming a woman that attracts a man to the point of wanting to "become" one? Or what is it about becoming a man that attracts a woman to the point of wanting to "become" one? At the same time, why is there an incompatibility between the attractive qualities of womanhood and remaining male? Or why is there an incompatibility between the attractive qualities of manhood and remaining female? In other words, is it not a part of normal development to aspire to possess or to manifest a quality which one admires in another?

But since when is a human person independent of being a female-person or a male-person? To be either male or female is a type of personhood; and, therefore, it is the type of personhood that defines the subject who undergoes

---

[36] Cf. the following article which cuts across the media tendency to the unquestioning acceptance of sweeping ideas. Abigail Shrier, "Standing Against Psychiatry's Crazes": an interview with Dr. Paul McHugh who, in 1979, closed the sex-change clinic at Johns Hopkins; and, as a part of the interview extract she quotes what he says: "*"Everybody should agree" that sex-reassignment surgery is "an experiment right now," he says. "We're doing an experiment. We have lots of publications that are telling us that the evidence base for these treatments is very low-quality"*": https://www.wsj.com/articles/standing-against-psychiatrys-crazes-11556920766?curator=MediaREDEF.

change. If, then, a person modifies his appearance so that he appears more feminine, he is still the same person he was before the modification: a person is a whole and not a partial aspect of human identity; and, therefore, a change in appearance is not a change in identity but a modification of an existing identity: a man looking more like a woman is still a man who looks more like a woman.

The whole human being-in-relationship: An incarnate expression of the interiority of the human person

A human reaction, however, bears an inwardness which makes the physiological response a kind of *incarnate expression* of the interiority of the person. On the one hand even the ordinary everyday activities of getting ready for the school run, filling bags, going in and out for children, cleaning the car windows and warming the car and then driving the children to school can be charged with a number of reactions ranging from being border-line late, hoping to get through the traffic lights before they turn red, playing I-Spy with the children in the car and trying to keep them guessing for the whole journey, thinking of clues and sending them off, happy, once we arrive. In other words, driving the children to school is about de-stressing the getting ready and enjoying the journey-time with my children; and, therefore, all the ingredients "embody" a mentality that works out in terms of many discreet but related activities. Travelling, too, is about that very specific time of being *en route* but, in the course of it, spending that time with children who have come to value that time with Dad; and, therefore, it is an opportunity to take an interest in the variety of everyday experiences that are a part of school, growing up, recreation, reading, films, friendships and considering the possibilities of life.

On the other hand, each one of these activities had a heart-rate, pulse level, nervous-endocrinological response, brain activity, breathing, motor activity, eye, ear, hand and foot coordination and many other aspects all of which fluctuated as the factors varied *en route*; and, even more specifically, there are the passing frustrations of a forgotten school bag, phone or trainers, the delight taken in the new growth on the roses, the pinks in the sky and the wit of a comment or other response from the children, thanking another driver, encouraging a person to cross the road and blessing God for getting through the traffic in time for school.

In the case of human beings, then, the physiologically expressed mentality of being a father or a mother is not transplantable; it is, as it were, a function of the whole being, masculine or feminine. This is not to deny the possibility of variations within the complementary differences appearing to overlap, as it were, the differences between the "biological" sexes[37]. Nevertheless, however, a man's femininity is different to a woman's femininity and a woman's masculinity is different to a man's masculinity. In other words, just as there are attitudinal echoes between the sexes so there are physical characteristics which appear in more or less of the other sex; but, even then, the physical characteristics are a part of a different physiological whole according to the person being male or female.

Just as there are anatomical differences between men and women, physiological differences which accord with anatomical differences, so there are psychologically inscribed differences of mentality which unfold the relational nature of human being; and, therefore, the pelvic structure of men and women are different in that the pelvic structures of a woman is wider and capable of child bearing, physiological changes in the woman express the welcome of the child which, psychologically, is expressed in expecting to meet the child conceived. Thus there is an ever more necessary understanding of the integral nature of being a man or a woman; and, indeed, of the wholesome beginning, unfolding and development of each.

---

[37] Cf. William J. Malone, Colin M. Wright and Julia D. Robertson, "No One Is Born in 'The Wrong Body'", September 24th, 2019, *Quillette*: https://quillette.com/2019/09/24/no-one-is-born-in-the-wrong-body/.

Marriage and Parenting: A Dynamic Conclusion

'In the family, children "learn to love inasmuch as they are unconditionally loved, they learn respect for others inasmuch as they are respected, they learn to know the face of God inasmuch as they receive a first revelation of it from a father and a mother full of attention in their regard"[38].

'[W]omen's 'capacity for the other' favours a more realistic and mature reading of evolving situations, so that "a sense and a respect for what is concrete develop in her, opposed to abstractions which are so often fatal for the existence of individuals and society"'[39].

'Women have a unique understanding of reality. They possess a capacity to endure adversity and "to keep life going even in extreme situations" and hold on "tenaciously to the future"[40]. This helps explain why "wherever the work of education is called for, we can note that women are ever ready and willing to give themselves generously to others,

---

[38] Congregation for the Doctrine of the Faith, Letter to Bishops "On the Collaboration of Men and Women in the Church and in the World", May 31st, 2004, quoted on p. 58 of the *YOUCAT* Catechism, translated by Michael J. Miller, London and San Francisco: Catholic Truth Society and Ignatius Press, 2011.

[39] "Male and Female He Created Them": Towards a Path of Dialogue on the Question of Gender Theory in Education, 17, quoting from the Congregation for the Doctrine of the Faith, Letter to Bishops of the Catholic Church on the Collaboration of Men and Women in the Church and in the World, 31 May 2004, 13.

[40] "Male and Female He Created Them": Towards a Path of Dialogue on the Question of Gender Theory in Education, 18, quoting from the Congregation for the Doctrine of the Faith, Letter to Bishops of the Catholic Church on the Collaboration of Men and Women in the Church and in the World, 13.

especially in serving the weakest and most defenceless. In this work they exhibit a kind of affective, cultural and spiritual motherhood which has inestimable value for the development of individuals and the future of society'[41].

'Saint Joseph offered the child growing up beside him the support of a healthy masculinity, a clear understanding of human problems, and courage …. For Saint Joseph, life with Jesus was a continuous discovery of his own vocation as a father. He became a father in an extraordinary way, without begetting his son in the flesh'[42].

Marriage, then, is a vocational work and entails that dialogue between appreciation and the rubbing of relational difficulties with each other; indeed, it may be that the recognition and reconciliation of hurts is one of the enduring signs of the sacramental presence of Christ in the marriage: constantly turning the water of suffering into the wine of gratitude and gladness (cf. Jn 2: 1-11). In other words, a forgiving mentality that permeates marriage may be more beneficial to the children than efficiency and all the other practical benefits of having parents. Fatherhood, then, as with motherhood, is primarily fulfilled by the person who genuinely loves the child; and, indeed, it is as much a goal as a reality to be lived. In other words, it is not as if fatherhood is a ready-made vocation; indeed, in my experience, there is a kind of conversion to fatherhood: a turning away from the inconvenient interruptions of the needs of children to a recognition, however clumsy, of how to be of help to them. There is a kind of oversight, too, which belongs to fatherhood and which is especially evident in the life of a good bishop[43]: an oversight that ranges throughout the whole axes of relationships

---

[41] "Male and Female He Created Them": Towards a Path of Dialogue on the Question of Gender Theory in Education, 18.

[42] St. John Paul II, *Rise, Let Us Be on Our Way*, p. 140.

[43] Cf. The whole book: St. John Paul II, *Rise, Let Us Be on Our Way*.

but also examines, accompanies and judges, in conjunction with his wife, the whole plethora of the practicalities of daily life in the family.

In sum, then, the child's embryonically begun[44] unfolding biological development implies an enabling psychological development which matures a profoundly physiologically expressed psychological mentality of being a man or a woman. Thus the psychological characteristic of "being with" is both present in the "woman" and, at the same time, manifest in the structural relationship of maternity: a maternity which needs the complementarity of the differently expressed relationship of the child to the father. Indeed, just as the mother's relationship is expressive of "being with" the child so the father's relationship "takes up" the enabling to live-in-relationship which begins with the mother and recognizes the myriad ways of collaborating which are involved in being "drawn into" the social structure that "draws out" the talents and training possibilities of the child. In the process of parenting, however, the equally different maternal and paternal contributions are a natural dynamic which, nevertheless, need the perfecting help of graced self-knowledge, openness to criticism and an appreciation of the reality of the other's perception of each child and the family as a whole. In a word if the woman's nature is a kind of nurturing earth the father's is a kind of introduction to gravity's benefits and dangers; however, each may do the other in a way that is helpfully different: the difference being expressive of both their relationship to one another and their complementary differences.

---

[44] "Male and Female He Created Them": Towards a Path of Dialogue on the Question of Gender Theory in Education, 24: 'the sexual difference between men and women ... can be demonstrated scientifically by such fields as genetics, endocrinology and neurology. From the point of view of genetics, male cells (which contain XY chromosomes) differ, from the very moment of conception, from female cells (with their XX chromosomes).'

Foreword to Chapter Seven

Leah Palmer

Introduction

Chapter 7 addresses a bioethical issue which has become prevalent in the modern world. The modern developments of *in vitro* fertilization and surrogacy parenting have been praised by the culture at large for their attempts to "solve" the issue of infertility and even the "problem" of discomfort in childbearing.

These modern developments are only symptoms of a deeply flawed, yet generally accepted, understanding of God's design of the human person. While this flawed understanding has become the ideology of our modern culture, it has roots within Christ's Church. Yes, even within the Church, misinformation, misunderstanding, and misrepresentation of the role that fertility and children has in the marital union continue to proliferate. It is here, among the faithful, which certain misunderstandings contribute to the neglect of various hidden crosses that affect many married couples today. These hidden crosses have the potential to prevent couples from conceiving children early on in marriage, or in the case of complete infertility, ever having children.

My heart has broken time and again on behalf of couples that experience these circumstances. I have heard and met Catholic married couples who believe that their marriage is not properly "blessed by God" without the gift of children, and some who even believe that having more children implies greater favor in the eyes of God. As a result, how many couples have felt that if they don't have children, their marriage won't survive? I have personally met a few. So we can see that even within the Church, a flawed understanding of God's plan for humanity can thrive.

Whether it seems to be the case or not, what the Church and her members proclaim has an effect on the culture at large. The Church has been working to deepen her understanding of God's design and plan for humanity since its foundation, and her work will never be done. This is in large part what the *Second Vatican Council* aimed to do when addressing the matters of marriage, family, and human sexuality. We, the faithful, can join the Church in her mission to look deeper into the root causes of these ethical catastrophes rather adopting a merely reactionary role when faced with a confused culture.

My goal here is to highlight a few important definitions and distinctions made by the *Second Vatican Council* concerning God's plan for marriage and family, and to briefly discuss the logical and ethical conclusions that flow from that deeper understanding. This discussion is important for informing what the faithful proclaim to the world, and for our approach to healing the culture at large.

The role of children in the light of the theology of the body

Pope Saint John Paul II's *Theology of the Body* speaks a great deal about conjugal spirituality and the role that fertility, openness to life, and the gift of children play in the conjugal union.[1] In his Wednesday audiences – the collection of which comprise the book *Theology of the Body* – Pope Saint John Paul II communicates that children are an outward, physical sign of an inward, spiritual fruitfulness that occurs in the conjugal union. Expounding on the Pope's treatment of the subject, Christopher Stravitsch explains that

---

[1] Pope Saint, John Paul II. *Man and Woman He Created Them: a Theology of the Body*. Boston, MA: Pauline Books & Media, 2006.

"parenthood brings about a unique fulfilment in [the spouses'] conjugal spirituality."[2] In other words, even when the spousal union does not produce a physical offspring, it still bears spiritual fruit. This spiritual union produced through the sexual intimacy of a married couple precedes physical offspring, and therefore contains an intrinsic value and blessing in and of itself.

As soon as two persons are joined in holy matrimony, they become a family. Whether or not a married couple have children, they have "started a family" when the man leaves his father and mother and cleaves to his wife.[3] A married couple who is unable to have children are no less a family than a couple with 3 children or a couple with 10 children.

The goods of marriage

The sacrament of marriage is a covenant "by which a man and a woman establish between themselves a partnership of the whole of life and which is ordered by its nature to *the good of the spouses and the procreation and education of offspring*"[4] [emphasis mine]. These two components that orient the marital covenant, collectively understood as "the goods of marriage", need to be properly understood.

Before the Second Vatican Council, the procreation of children was considered the primary "good" of marriage, while the "good" of the spouses was considered secondary. With the Second Vatican Council, however, the Church came to a deeper understanding of God's plan for marriage, more carefully considering the needs that vary with every couple in light of the "goods" of marriage. The 1983 *Code of Canon Law* and its subsequent

---

[2] Stravitsch, Christopher J. "Spousal Love in Conjugal Spirituality." Homiletic & Pastoral Review, September 24, 2014. https://www.hprweb.com/2012/09/spousal-love-in-conjugal-spirituality-2/.

[3] Eph. 5:31 (ASV).

[4] *Code of Canon Law*, c. 1055, § 1, in *Code of Canon Law: Latin-English Edition*. Washington, DC: Canon Law Society of America, 1999.

Catechism expound on this deeper understanding, treating the "good of the spouses" as equal to that of procreation. Let's look for a minute at what this small but mighty distinction means for marriage and family.

When the good of procreation is placed before the good of the spouses, the dictum for married couples too easily becomes "as long as there are children, it is acceptable to neglect having an equal partnership, mutual love, and caring for each other's needs." This clearly does not reflect the love that Christ has for His church, which marriage is meant to sacredly represent.

Further down this miserable rabbit hole of consequences, when we place the good of the spouses and the spiritual fruitfulness of the conjugal union second to the good of procreation, we neglect to proclaim to the world the full purpose of human sexuality. When we understand the good of the spouses and their spiritual fruitfulness as necessary for procreation, the modern "option" of *in vitro* fertilization and surrogacy parenting are unacceptable simply because it excludes this precedent. Children are intrinsically good, but we must be careful not to risk emphasizing the good of children at the cost of the good of the spouses.

Human biology and God's plan for co-creation

With a better understanding of the goods of marriage, we ought to spend some time deepening our understanding of what God's design of human biology can tell us about accomplishing the balance of those goods. Indeed, our biological design plays a role in assuring that the good of the spouses is not neglected within marriage. A deeper understanding of our biology and what Pope Saint John Paul II referred to as "responsible parenting"[5] ought to be incorporated in discussions surrounding this topic. The inability to have children" encompasses many situations, traditionally referring to only cases

---

[5] Pope Saint, John Paul II. *Man and Woman He Created Them: a Theology of the Body*. Boston, MA: Pauline Books & Media, 2006.

of natural infertility. But many couples have difficulties unrelated to infertility that still prevent them from having children. This is where the concept of "responsible parenting" can be applied in a useful way.

The concept of "responsible parenting" encourages and allows space for couples to fully participate in the co-creative power that God desired for human sexuality. "Responsible parenting" acknowledges a relationship with God in which He calls married couples to be co-creators with Him by "taking into account their physical and psychological health, their existing duties and responsibilities, especially other children, as well as their financial circumstances in the decision to have children. Practically speaking, this means prayerfully approaching the decision, using your reason, respecting the gift of your fertility and considering the various obligations you already have."[6] This relationship avoids both extremes of "deciding without God" as well as the idea that "God decides without us."

Much traditional thought around natural family planning[7] still reflects a time when procreation was considered the primary good of marriage, before our deeper understanding of the conjugal union was developed by the Second Vatican Council. With a now deeper understanding of God's intention for this union, the Catechism of the Catholic Church clearly states: "A particular aspect of this [paternal] responsibility concerns the regulation of births. For just reasons, the spouses can desire to space the births of their children. It is up to them to ensure that their desire does not depend upon egoism but is conformed to the right generosity of a responsible paternity."[8]

---

[6] Archdiocese of Denver. "Theology of the Body and Natural Family Planning." n.d. https://www.archden.org/archbishops_writing/theology-body-natural-family-planning/#.Xf4gsehKjIU.

[7] Society of Saint Pius X. "The Problem of Natural Family Planning," January 8, 2016. https://sspx.org/en/news-events/news/problem-natural-family-planning-3180.

[8] *Catechism of the Catholic Church.* Vatican City: Libreria Editrice Vaticana, 2019.

Couples who are "unable to have children" but aren't particularly infertile might be, for example, the mother who, during her first pregnancy, experienced certain physical or psychological traumas that left her living in a state of post-traumatic stress for several years after the birth, or the couple who is experiencing the death of a family member whose funeral expenses has left them in crippling debt. In their desire to delay pregnancy due to their physical, psychological, financial, etc., burdens, they should not be excluded from our consideration because they, too, experience a particular aspect of "the inability to have children" by way of choosing to be responsible with their gift of fertility.

We must strike a fine balance here, neither over-emphasizing the good of the spouses nor the good of procreation, but always keeping in mind the major bioethical consequences that an imbalanced understanding of the two can have on the broader culture. On one hand, emphasizing the good of the spouses at the cost of the good of children produces an environment where man-made contraceptives and abortion are seen as a "solution" to abundant fertility. Yet on the other hand, placing the good of procreation before the good of the spouses produces an environment where in vitro fertilization and surrogacy parenting are seen as "solutions" to infertility. We must keep a keen focus on God's design and plan for humanity, continue to deepen our understanding of this design and plan, and continue to seek new and effective routes for proclaiming it to the culture at large.

CHAPTER SEVEN

LOVE, SCRIPTURE, SUFFERING AND BIOETHICAL QUESTIONS

General Introduction to Chapter Seven: The Challenge to Believe God is for us! Sufferings are so brutal that they dismantle us and tempt us to believe that God does not exist or does not even know about us - never mind loves us! Or that if there is a God who witnesses the misery of our lives and does nothing – How terrible if that were true! Sufferings bring us to death: to want to die and to be done with a life so unbearably unacceptable to us! Sufferings are a language unintelligible to us: Does not every ounce of human goodness say – "If I were God would I let human freedom be as free as God allows?!"

At the same time, in terms of the difficulties encountered in the culture in which we live, there are three problems which imply a breakdown in the possibility of understanding the value and mystery of marriage: the availability and use of pornography, suggesting a blindness to the existence of the person as an 'incarnate spirit' (*Familiaris Consortio*, 11); the temporary 'hookup' culture which seems to substitute sex for conversational intimacy; and, thirdly, the cycle of divorce and its impact on the possibility of believing in marriage as being husband and wife, life-long and open to life[1].

Add to this the mentality that everything and everyone is "plastic", that every problem requires a technological intervention and that there are no principles, only arguments that more or less appeal to human sympathy; and,

---

[1] Randall Woodard, "Our Current Youth Culture and its Upcoming Impact on Successful Marriages", https://www.hprweb.com/2017/11/our-current-youth-culture-and-its-upcoming-impact-on-successful-marriages/.

therefore, *in vitro* fertilization, appealing to the plight of infertile couples, does not actually offer a remedy for infertility but, rather, seizes the opportunity to intervene, technologically, as if the problem does not involve the whole human being: the whole human being of the man, woman or child.

Thus there is nothing neutral about the "situation" of today and its many facets: facets that are actually profoundly implicated in the deepest questions of the human heart, the experience of agonizing sufferings and the problem of being unable to reason to and from the reality of the whole human being-in-relationship. The response of love, truth and goodness to all the difficulties of life has to begin again, then, in the new context of the regeneration of human life and culture, drawing on all that is good, true and loving in the practices of "today"; but, at the same time, this new beginning is not wholly new but is integral to the new beginning which God is constantly bringing about and which we are continually rediscovering. In the literature drawn from the Early Fathers of the Church it is possible to come upon unexpected thoughts in the writing of the Church Fathers and, as we increasingly recognize, the great Mother of the Church. Thus, in an excerpt from the work of St. Cyril of Alexandria, (d. 444), we read that 'Many notable things were accomplished in this one sign [when Christ turned water into wine at the marriage feast of Cana], his first sign. Honourable marriage is sanctified, and the curse pronounced against woman is overcome. Women will no longer bear children in sorrow, since Christ has blessed the very beginning of our lives'[2]. Thoughts that St. John Henry Newman continues, both recognizing how 'polygamy and divorce' were 'detrimental to the dignity of women'[3] and asserting that 'women "shall be saved through the Child-bearing, that is, through the birth of Christ from Mary, which was a

---

[2] Footnote 7: *Commentary on John* 2, 1; PG 73, 228 cited on p. 245 of *Mary and the Fathers of the Church: The Blessed Virgin Mary in Patristic Thought*, by Luigi Gambero, translated by Thomas Buffer, San Francisco: Ignatius Press, 1999.

[3] From the commentary by Philip Boyce, p. 71 of *Mary: The Virgin Mary in the Life and Writings of John Henry Newman*.

blessing, as on all mankind, so peculiarly upon the woman'"[4].

Chapter Seven: Introduction: The Word of God and Human Experience. It might seem that in an age when there is an almost instant transmission of information around the world, the production of pictures of both distant planets and the microscopic moments of human conception and the detailed examination of the flooding of the seas, the trapping of animals or the congestion of their intestines with plastic - that we live in a world *beyond the reach of the word of God*. Human experience, however, is invaluably expressed in a word which has originated through the age-old experience and sufferings of mankind. The word of God goes back to the very origin of the act of creation, magnificently accounting for the wondrous existence of what did not exist before it was called into existence and then, through the intricate and intimate experience of men and women, traversing the mysteries of sin and redemption that witness, throughout a wealth of human relationships, situations, events, attitudes and acts of God. Even if, then, there is not the precise account of the contemporary "problem" there is its earliest manifestation, roots and account of its pain and the healing which comes through a dialogue with God

On the one hand there is the primordially "fallen" desire to reject the possibility of love's limits: to reject the possibility that human being entails within it a compass which points to the exploitation, abuse or destruction of human life; indeed, that just as human being is claimed to be infinitely plastic so the justification of everything is claimed to be possible by juxtaposing a true good with an objectively unrelated goal. Thus the objectively existing suffering of the 'blocked fallopian tubes' of a wife who wanted a child, whose tubes that direct the egg from the ovary to the womb, meeting *en route* the fertilizing sperm, were blocked; and, therefore, the objective suffering of this

---

[4] Quoted from *Parochial and Plain Sermons*, Vo. II, p. 131 in *Mary: The Virgin Mary in the Life and Writings of John Henry Newman*, p. 71.

woman and indeed of her husband became the pretext for a completely unrelated procedure of 'combining a man's sperm and a woman's ovum in a glass dish'[5]. Clearly *in vitro* fertilization is not a cure for a woman's blocked fallopian tubes. Thus it is that many unethical procedures are aligned with an unrelated good to justify them by the "glow" of what is truly good. It is indeed good to search for a cure for infertility, whether that cure is a remedy to the cause of blocked fallopian tubes, the actual blockage in the fallopian tubes or some other cause of male or female infertility.

On the other hand, then, Scripture not only addresses the experience of suffering infertility it begins, holistically, with recognizing the very origin of the desire to procreate originating with the original gift of the Creator: 'And God blessed … [Adam and Eve], and God said to them, "Be fruitful and multiply, and fill the earth and subdue it; and have dominion over the fish of the sea and over the birds of the air and over every living thing that moves upon the earth"' (Gn 1: 28). Let us begin, therefore, with the command to "Be fruitful and multiply" and the Cross of Infertility (I) and then go on to the biblical account of the suffering of infertility (II) ending up with Love and Bioethics (III).

The Command to "Be Fruitful and Multiply" and the Cross of Infertility (I)

We live in a time when the blessing of fertility has, for many in positions of ideological power, power driven by an incomplete account of reality, ceased to be a blessing and become a curse; indeed, even if each one of us has been given the gift of life, the tragic irony is that there are those who, having received this gift, seek to eradicate it from the face of the earth. In other words, the blessing of fertility has become in the eyes of a powerful elite a curse for the welfare of the earth; and, indeed, all the chemicals poured into

---

[5] Etheredge, *The Human Person: A Bioethical Word*, p. 350.

the sea which arise out of the commercial expression of the suppression of human fertility are deforming or interfering with the fertility of other creatures: polluting the earth and compounding the problems of the poorest of the poor. The tragic nature of this faithless fear (cf. *Gaudium et Spes*, 36) of overpopulation, this obsessive preoccupation with destroying the gift of fertility and this carelessness of the lives and living conditions of people throughout the world, makes it necessary to look at the 'good' of fertility which has become so disfigured, maligned and demonized: the good of human life; indeed, if all the medical expertise and experiments on human embryos were really about the good of human life then not a single life would be lost – therefore it is clear that there is a different motive to that of the love of life to which all the aborted, neglected, frozen, discarded and experimented-upon human embryos have tragically witnessed.

"Be Fruitful and Multiply"

Even before there is a consciousness of God there is an admiration in front of the wonder of a new born baby[6], a natural delight in the magnificent presence of the mystery of the child, a desire to acknowledge and celebrate the woman who has given birth (cf. Tobit, 4: 3-4) and the father who loves the unfolding of the spousal love between himself and his wife. There is, in other words, a deeply "ingrained" openness to life which speaks, as it were, of a meaning which transcends "reproduction" and translates "procreation" as the participation in the very creation of a human being: a human being

---

[6] In the context of a profoundly anti-life mentality, why would a child not say, on seeing a pregnant woman: 'What's wrong with that Fat Lady?' Not to mention the whole definition of pregnancy as a depersonalized process: 'Pregnancy is the temporary deformation of the body of the individual for the sake of the species' (Firestone, p. 180 of *The Dialectic of Sex: The Case for Feminist Revolution*). Indeed, where is the sensitivity, anyway, to a woman being called a 'Fat Lady' and to the fact that each of us was a child carried by a mother?

who is always a human being-in-relationship – just as being conceived is always and only "to be conceived-in-relationship". In other words, like man and woman, husband and wife, mother and father, a *child* is an expression of "relationship": of bearing the likeness of the human family (cf. Gn. 5: 3) and transmitting what was both received from the parents and goes beyond them in the originality of being given existence.

On the one hand it is possible to speak of 'man naturally ... [seeking] whatever accords with his generic animal nature, whatever *nature teaches all animals*: mating between the sexes and bringing up one's young'; however, this is distinguished from the fact that 'man naturally seeks whatever accords with ... [his] rational nature ... to know the truth about God ... and to live a social life'[7]. In other words, while it is possible to understand the comparison between human beings and all animals from the point of view of 'whatever *nature teaches*', the problem is that this does not begin to express the personalistic meaning which derives from understanding man and woman to be created in the image and likeness of God, seeking a relationship to the truth and living socially because we are, fundamentally, beings-in-relationship. Indeed, it could be argued, nature teaches males to be promiscuous, uninterested in the young and appetite-driven; and, therefore, there needs to be an integration of seeking the truth about God, marriage and parenthood, all of which unfold together in love, through love and for love.

On the other hand, then, if we are made in the image and likeness of God, and God created all that exists from nothing and loves everything that He created with an eternal love, then it follows that the command to be 'fruitful and multiply' (Gn. 1: 28) exists in relation to both being created in 'our image, after our likeness' (Gn. 1: 26) and in view of the vocation to 'fill the earth and subdue it' (Gn. 1: 28). In other words, there are three interrelated strands to this beautiful, procreative "chord": to be created in the image and

---

[7] St. Thomas Aquinas, *Summa Theologiae*, Methuen, p. 287: I-II, Qu 94, Art 2.

likeness of the Blessed Trinity; procreation-as-collaboration with God; and, thirdly, drawing on all that is good, cultivating and developing the earth. The deepest origin, then, to the nature of procreation is both the mystery of God and His providence; and, albeit the focus here is on the mystery of God, on the "presence" of the God in the mystery of spousal procreation. The bare mystery of God creating from nothing all that exists (cf. CCC, 296-298) is, as it were, clothed in the reality of human experience: that just as creation comes to exist from nothing so each person that comes to exist comes to exist from nothing. Recognizing, however, that whereas the absolute beginning of creation entailed the coming-into-existence of all that exists except the being of God who "Is" (cf. CCC, 205-214), there is nevertheless more than an echo of recapitulating the original act of creation in the coming-to-exist of each one of us. Through procreation, therefore, the parents are truly co-creators with God of a human being who did not hitherto exist; and, as such, the very originality of creating "from nothing" must in some sense permeate the fruitfulness of spousal love.

Spousal love is both "we" and the possibility of being parents of a child

When Adam and Eve become 'one flesh' (Gn. 2: 24) they "imitate", as it were, the unity of God in being creative; and, therefore, there is inscribed within the nature of procreation an inherent intimation, as it were, of the mystery of a human person "coming forth" – *not from nothing – but from the creative action of God.* In the reality of procreative spousal love, in other words, there is as it were an "enfleshed" expression of a human being coming forth from the creative act of God; and, in terms of temporal time, there is a relationship between the "communion" of spousal love's enfleshed echoing of the divine "We" and the coming forth of a human being from the depths of a mysteriously generative love. Thus the unity expressed in Adam and Eve being 'one flesh' is a unity-in-communion; and, as such, does not absorb but entails, encompasses and expresses the diversity of being male and female, husband and wife and, potentially, father and mother. When, therefore, the mother of seven martyrs spoke about the mystery of procreation that they might have the hope of the resurrection, she said 'I do not know how you came into being in my womb' (2 Mac. 7: 22) – such that she was referring to the deepest mystery of her children's existence transcending what it is possible for human beings to do; and, if this is not clear, she also goes on to suggests more explicitly that this mystery is involved in the coming-to-be of human beings: 'look at the heaven and the earth and see everything that is in them, and recognize that God did not make them out of things that existed. Thus also mankind comes into being' (2 Mac. 7: 28).

There is, then, the twofold word of both communion and fruitfulness; and, therefore, the desire to be fruitful must be ordered to spousal communion. In other words it is deeply embedded, even enfleshed in the physio-psychological structure of spousal communion, that there is a possibility of a child issuing from their intimate embrace of each other; but, at the same time, the intimate embrace both precedes and accompanies fruitful spousal love – even when it is not fruitful. Therefore there is in spousal communion a word which is an intelligible expression of love being about "being-with" each other: a being-with which precedes, whether or not love bears the fruit of children, accompanies and supersedes the coming of children if indeed they came. Parenthood, then, is a kind of "multiplication" of the being of spousal love; and, in the words of St. John Paul II, spousal love both is and generates a "communion of persons" (*Familiaris Consortio*, 15). Clearly, spousal love is complementary to but different from parental love; and both loves are, as it were, ordered to the wider love of all people that begins to be expressed in the mission of the family to others.

The inner-cross of infertility: being unable to beget in the begetting act of God

If, then, there is a dynamic within the very nature of spousal communion which shows itself in the being-with of the husband and wife's togetherness, originally expressed in the very choice of each other in the gift of marriage, then there is a profound swelling, as it were, of that love as it is poured out in marriage: as it almost seeks to replicate itself in the increase of the communion of persons with which it began. Just, then, as the "being-together" of the Blessed Trinity expresses an enraptured abundance of "being" three persons in one God, so there is a kind of enraptured overflowing of the being together in marriage which seeks, as it were, an embodied expression in an increase in the communion of persons in the family.

The cross of infertility, then, is not an imaginary, superfluous or an extraneous experience; rather, it strikes at the very core of the abundant budding of love: that love is generative of personhood. Just as in the mystery of the Blessed Trinity the existence of the three persons in one God is inseparable from God being Love, so human love is an enfleshed expression in time of the eternally being Love of the Father, of the eternally being generated Love of the Son and the eternally proceeding Love of the Holy Spirit from the Father and the Son. Procreation, then, is the enfleshed expression of love-begetting-personhood. In that the begetting act of God brings "human personhood" and parenthood to be from the being-together of spousal love – then it follows that the frustration of this inner dynamic is fundamental to the suffering of infertility. The command to be 'fruitful and multiply' is really an expression of the inner-nature and dynamic of love; it is not, in other words, an imposition from without it is, rather, an embodied expression of the inner nature and mystery of the Blessed Trinity.

The Biblical Expression of the Experience of Infertility (II)

The Bible as a whole recounts many and varied experiences of the suffering of infertility; and, in general, this often entails an appeal to God to act: an action of God that culminates in the mystery of the *Incarnation* being wholly beyond the horizon of human experience and the deepest expression of God's "enfleshed" manifestation of His own mystery. Thus the virginally chaste love of Mary and Joseph is the spousal love which expresses, as deeply as it is humanly possible, the virginally chaste love that is God. This virginally chaste spousal love is the fitting human expression of the Love that God "Is", entailing as it does, the begetting of the person: the person of Jesus Christ – true man and true God – in and through and from the virginally chaste love of God expressed in the Motherhood of Mary. In other words, the chaste spousal love between Joseph and Mary is not irrelevant to the mystery of God's creative love: their chaste spousal love is a sign, almost a presence, of the mystery of Love; therefore in a remarkably discreet way, almost silently, the spousal love of Mary and Joseph shows forth the splendour of a love that "imitates" the completely self-less Love of the Blessed Trinity. This Love that is God is a Love which, in the words of St. John's Gospel, is a uniquely self-less and self-giving Love such that each person is loved into existence and conceived 'not of blood nor of the will of the flesh nor of the will of man, but of God' (Jn. 1: 13)[8]: a Love that is utterly free and unfettered by any kind of necessity: a Love that loves the person as completely as it is possible to Love (cf. *Gaudium et Spes*, 24).

However imperfect human love is there is, as it were, a spark of the utterly self-less Love that God expresses in the conception of each human being: the Love that treasures each person as 'the only creature on earth that God has

---

[8] While it is true that the context of this verse is the mystery of those to whom 'he gave power to become children of God' (Jn. 1: 12); nevertheless, the verse seems to speak of a completely free, generous and self-less love – the Love that God "Is".

willed for its own sake' (*Gaudium et Spes*, 24); and, therefore, it is possible to glimpse how deeply embodied in human love is the possibility of expressing the Love that God expresses in bringing a human being to exist. In other words, the Love that God expresses in bringing each man and woman to exist, is a Love that is eternally irrevocable; and, therefore, even before a particular person is discovered to be infertile, there is the reality of being Loved into existence by God – for his or her 'own sake' (*Gaudium et Spes*, 24). Believing-to-be-loved is the foundation of each person's existence. In front of infertility, then, there is a temptation, however unconscious, to consider the blessing of being loved-as-blighted, as it were, because it looks as if the gift of being given life has been given incompletely; and, indeed, if the Church has traditionally seen a 'large family [as] a sign of God's blessing' (CCC, 2373) – it is possible that infertile people experience themselves as "cursed", unworthy·or even unloved.

In the following, closer scrutiny of the word of God, we see a whole range of human reactions to the suffering of infertility. Indeed, there is a whole "mess" which is both a part of marriage and goes beyond it; and, at the same time, just as the birth of any child is always an act of God, so God is always bringing good out of the difficulties that these men and women experienced. At the same time, even if there is not the technological involvement in the transmission of life, there is a whole complex of relationships which act, in a way, as the equivalent to the modern-day experience of "sperm donor", "egg donor" and "surrogate mother".

Three generations of infertility: Abraham and Sarah; Isaac and Rebecca; and Jacob and Rachel

'And [God] … said [to Moses], "I am the God of your father, the God of Abraham, the God of Isaac, and the God of Jacob"' (Exodus, 3: 6).

The first generation of infertility: Abram, Abraham and Sarai, Sarah

Even if, then, husband and wife are surrounded by wealth and all kinds of benefits the experience of infertility, which is really an experience of pain in the depths of marriage, is an experience which almost defines a person. Abraham, who had been encountered by God, was promised that if he went from his country He would make of him "a great nation …. and by … [him] all the families of the earth shall bless themselves" (Gn 12: 2-3; cf. also Gn 13: 15); and, therefore, this is a promise in the face of his childlessness, to which he adverts when he speaks of his pain: 'And Abram said, "Behold, thou hast given me no offspring; and a slave born in my house will be my heir"' (Gn. 15: 3) – to which God replies "[Y]our own son shall be your heir" (Gn. 15: 4).

Sarai, Abram's wife, sees the possibility of Hagar, her 'Egyptian maid' (Gn. 16: 1), going in to Abraham and giving Sarai a child by a kind of surrogacy; however, although Hagar bears Ishmael, the relationship changes between Sarai and Hagar: Hagar 'looked with contempt on her mistress' (Gn. 16: 4) and 'Sarai dealt harshly with her' (Gn. 16: 6). Later, God makes it plain that Sarai, renamed Sarah, will be the mother of Abram's child (Gn. 18: 9-15) and Abram is also renamed and is now called Abraham. But when Isaac is born to Sarah and Abraham, Sarah wants Hagar and Ishmael to go, although it 'displeased' Abraham (Gn. 21: 12); and, in a certain way, the word of God distinguishes between Sarah, Abraham's wife and 'the son of the slave woman' (Gn. 21: 13) – although both Isaac (Gn. 21: 10) and Ishmael (Gn. 21: 13) are recognized as Abraham's sons.

Without digressing too much, then, it is clear that although the son promised to Abraham was also to be born of Sarah, they nevertheless went through a period of thinking that they had to make what God promised happen through their own plan; however, the birth of Isaac seemed to inspire the painful realization that Abraham and Sarah had been mistaken and that God had intended to heal the infertility in their marriage, respecting as He did the fact that Abraham and Sarah were husband and wife. We see, also, Abraham and Hagar's suffering in view of having conceived Ishmael; and, therefore, there are already the nascent problems which will emerge as surrogacy[9] arises in modern times: the fact that "relationships" are more integral to marriage and family life than people realize. At the same time, however, we see that both Sarah and Abraham, for a while, "imagined" that "they" could have a child by Hagar and, in that way, overcome her own barrenness; and, therefore, we see the depth of Abraham and Sarah's desire for a child obscuring the intimacy of their marriage being exclusive to them and introducing conflicts and tensions which go beyond the suffering of infertility itself.

More widely, then, even if there are multiple possibilities that are technologically possible, the underlying reality of marriage and the parental relationships which arise out of conceiving a child, have a dynamic which unfolds an identity from the depths of the husband and wife's reciprocal gift of self (cf. *Gaudium et Spes*, 48-50). This family identity includes the possibility of adoption and charitable works but it is not divisible into

---

[9] Cf. "Broken Bonds: International Surrogacy Conference + Australian Book Launch": http://www.stopsurrogacynow.com/broken-bonds-conference-book-launch/?mc_cid=903fa1364c&mc_eid=9054b087ff#sthash.uXvUrzTu.va58nemo.dpbs.

As regards the book launch: 'In this book [Broken Bonds: Surrogate Mothers Speak Out], courageous women from the USA, the UK, Canada, Australia, India, Austria and Russia share their stories of becoming 'surrogate' mothers out of kindness and compassion (or need for money), only to be deceived, neglected, abused, harassed, or abandoned by 'baby buyers', clinics, and lawyers. Their stories are tragic, shocking, and revelatory of a profit-driven industry that preys on desperation and women's compassion': https://www.amazon.com/Broken-Bonds-Surrogate-Mothers-Speak/dp/1925581551/ref=sr_1_1?ie=UTF8&qid=1550605634&sr=8-1&keywords=Broken+Bonds%3A+Surrogate+Mothers+Speak+Out

"biological" relationships in which there are parental parts, as if one parent can "substitute" a spouse for another or both can substitute themselves for "others"; rather, husband and wife are an integral whole which speaks of a depth which arises out of marriage identifying husband and wife in a new way: an ontologically new way which is especially evident in the sacrament of marriage: an outward sign of consent to a new nature to their relationship. It is as if, in other words, each marriage constitutes a radically new beginning which, as it were, recapitulates "the beginning": the beginning which was "begotten" of the mystery of God the Blessed Trinity.

The second generation of infertility: Isaac and Rebekah

When Abraham's servant discerned that Rebekah was an answer to his prayer to find a wife for his master's son Isaac and, on the basis of her agreement to this (cf. Gn. 24: 58)[10] they were about to leave, it seems that Rebekah's mother and brother bless her: 'And they blessed Rebekah, and said to her, "Our sister, be the mother of thousands and ten thousands; and may your descendants possess the gate of those who hate them!"' (Gn. 24: 60). Naturally, in one sense, it is possible that this blessing was a part of the Jewish tradition that expressed the good expectation of children; however, in

---

[10] The question of Rebekah's freedom to marry Isaac does not seem to be simply a matter of it being determined by custom (cf. p. 128 of *The Navarre Bible: The Pentateuch*, translated by Michael Adams, Dublin and Princeton: Four Courts Press and Scepter Publishers, 1999) because, on the one hand, there is her willingness to go and the whole account of Isaac and Rebekah's love of each other and, on the other, there is a discernment of the will of God of which Abraham's servant has spoken and on the basis of which he has acted. When my own mother first saw my father, before she had even spoken to him, she experienced the sense that this man would be involved in the rest of her life and, indeed, they married and had seven children. Did my mother's sense that she was about to meet the man she would marry "take away" from him his freedom to marry – their freedom to marry?

Rebekah's case, it turned out that she was barren and, therefore, there was the whole experience of a long period in which there were no children.

On the one hand it could be construed that the contrast between the blessing and her barrenness intensified the suffering of their infertility, although there is no mention of this; rather, the text simply understates the fact that 'Isaac prayed to the Lord for his wife, because she was barren; and the Lord granted his prayer, and Rebekah his wife conceived' (Gn. 25: 21). The text understates this period of prayer on the basis that, in reality, the length of time between their marriage and the birth of twins was in fact twenty years; and, therefore, as the commentator says, Isaac 'spent twenty years praying'[11] and St. John Chrysostom remarks on Isaac's perseverance in contrast to how we can 'show some little zeal, … [and] lose heart and fall away if the response is not immediate'[12]. In other words, God's answer to prayer is a part of the history of salvation and not just a simple matter of responding to a request, because it also involves trusting in the promise to Abraham that God will make of him 'a great nation' (Gn. 12: 2); and, therefore, in the concrete circumstances of our lives there is often the invitation, through the facts of life, to persevere in prayer as well as the specific answers which, as it were, mark the path of our life and encourage us to continue.

On the other hand, then, the fact that 'Jewish and Christian tradition see the marriage of Isaac and Rebekah as a model of conjugal love'[13] takes up all that is characteristic of the actual relationship between them and indeed the particular circumstances of it: Isaac's trust in the discernment of his father's servant of the will of God, the blessing of abundant fertility upon Rebekah and her willingness to marry Isaac, whom she had not met, her barrenness, the "inheritance" of God's promise to Isaac's father, Abraham, and Isaac's

---

[11] *The Navarre Bible: The Pentateuch,* p. 133.

[12] *The Navarre Bible: The Pentateuch,* p. 134: excerpt from *Homiliae in Genesim,* 49, 1.

[13] *The Navarre Bible: The Pentateuch,* p. 129.

own patient, persevering prayer. There is, too, the testimony of Rebekah's turning to the Lord in the difficulties of being pregnant with twins: 'The children struggled together within her; and she said, "If it is thus, why do I live?" So she went to inquire of the Lord. And the Lord said to her, "Two nations are in your womb ..." (Gn. 25: 22-23); and, therefore, Rebekah was very much in the dialogue about the history of her life and how it was "in" the history of salvation into which she had married. Indeed, this perspective of our life and its difficulties as being integrally meaningful is an essential help in the day to day living of it.

The third generation of infertility: Jacob and Rachel

In this altogether complex account of the life of Jacob and Rachel there is a central dialogue which directly addresses the subject of infertility. Jacob was deceived into marrying Leah (Gn. 29: 25), the older of Laban's two daughters, and then went on to marry Rachel as well, who it was originally his intention to marry. According to the text, it says that 'When the Lord saw that Leah was hated, he opened her womb; but Rachel was barren' (Gn. 29: 31); but it does not tell us explicitly if she was hated by Jacob, Rachel or both of them, rather it emphasises the compassion of the Lord towards Leah. Independently of the circumstances of their life, it appears that the Lord responds to each of the two women in a way that is about their salvation; and, as such, the Lord is not prejudiced by the circumstances of their marriage – but rather looks at the heart of each (cf. 1 Sam. 16: 7). This then is the context of the ensuing dialogue between Jacob and Rachel.

'When Rachel saw that she bore Jacob no children, she envied her sister; and she said to Jacob, "Give me children, or I shall die!" Jacob's anger was kindled against Rachel, and he said, "Am I in the place of God, who has withheld from you the fruit of the womb?"' (Gn. 30: 1-2).

The rivalry between Rachel and Leah intensifies with each of them sending their maids into Jacob with a view to the competition as to who can "bear" more children; and, indeed, both children and maids almost cease to matter as they are subsumed under the consuming desire of the women to increase the number of births. Thus the disorder introduced by Laban deceiving Jacob, by giving him Leah instead of Rachel on his wedding night, grows into a multitude of difficulties and shows the tendency, already present, to be influenced by pagan customs: 'This sort of custom is explicable in a social context in which the slave-girl belonged to her mistress and therefore could be used by the mistress even as a surrogate mother'[14]. God, in leading his people, 'starts off by taking things as he finds them and then gradually trains the chosen people to appreciate human dignity and the true nature of marriage and the family'[15]. In other words, this is the possible starting point for all of us: God 'taking things as he finds and then gradually' training a person 'to appreciate human dignity and the true nature of marriage and the family'; and, therefore, whether we are single, married, divorced, separated, pregnant, a surrogate mother, infertile or fertile, God takes us from where we are and trains us 'to appreciate human dignity and the true nature of marriage and the family.'

"'[T]here are couples to whom our Lord does not grant any children. If this happens, it is a sign that he is asking them to go on loving each other with the same affection and to put their efforts, if they can, into serving and working the good of other souls" (Bl. J. Escriva, *Christ Is Passing By*, 27)'[16].

The ultimate answer, then, to the experience of infertility is in the discernment of the will of God: a discernment that does not necessarily

---

[14] *The Navarre Bible: The Pentateuch,* p. 152.

[15] *The Navarre Bible: The Pentateuch,* pp. 152-153.

[16] As quoted in *The Navarre Bible: The Pentateuch,* on pp. 155.

reject the possibilities offered by medical interventions; however, medical interventions are called to respect the inseparable connection between the unitive and the procreative significance of marital love: 'Only respect for the link between the meanings of the conjugal act and respect for the unity of the human being make possible procreation in conformity with the dignity of the person'[17]. In other words there cannot be a true ethical advance that does not recognize and respect the full dignity of both men and women and the integral requirements of each, unborn or growing up, whether single or married, whether poor or not.

Nevertheless the reality of the account of three generations of infertility makes one realize that *the word of God* is a place to find understanding and sympathy; indeed, in a certain sense, it is possible to say that the Lord is with the suffering of each person in all the particulars of it: He is with each one of us in the intimacy of His word. Therefore it is possible to say that the Lord understands each suffering in a way that is uniquely helpful – turn to His word and His Church for help!

Indeed, in a variety of surprising ways, the Lord's word speaks of the untold help of almsgiving, along with prayer and fasting: 'Store up almsgiving in your treasury and it will rescue you from all affliction' (*RSV*, Sirach, 29: 12); or, as it says in a different translation: 'Deposit generosity in your storerooms and it will release you from every misfortune' (*NJB*, Sirach, 29: 12). In other words, whether there is a cure or a change of heart about the suffering we encounter, or both, there is an indispensable help in the Lord through being generous to others; and, indeed, lest there be any doubt about the method of generosity, let it be anonymous or at least quietly discreet: "Thus, when you give alms …. do not let your left hand know what your right hand is doing, so that your alms may be in secret; and your Father who sees

---

[17] CCC, 2377, quoting CDF, *Donum Vitae*, II, 4.

in secret will reward you" (Mt. 6: 2-4)[18].

Love and Bioethics (III)

Suffering infertility expresses, acutely, the pain spoken of at the beginning of this book: the profoundly personal pain at the heart of human disappointment – but being disappointed in what? Ultimately, the answer is love. Whether parents are too rigid or unhappy to love their children[19], whether there is the death of a spouse[20] or one abandons the other, or children do not visit their parents or the suffering of infertility – the problem of pain is the problem of love.

Firestone, through all the difficulties she experienced, researched and wrote about, touches on love – love in the midst of its distortions and failures, albeit without any understanding of how it can be corrupted by anything other than power and 'inequalities'[21] – but a glimmer of shimmering love nevertheless:

'the self attempts to enrich itself through the absorption of another being.

---

[18] Cf. also the prayers of Hannah (1 Sam. 1: 12-20), wife of Elkanah, and John the husband of Elizabeth (Lk. 1: 13).

[19] A child can educate the father when she asks him: "Do you love me?" For, in the preoccupation with discipline and the fear of not being able to pass to the child what has helped the father, the father can lose the perception of the child's experience; and, instead of being preoccupied with whether or not a child is obedient, there is the deeper question of what help does the child need to see the good the father wants to bring about in the child's life: a good betrayed by the inability of the father to see the child's reality clearly.

[20] Cf. *A Grief Observed*, by C. S. Lewis, and Sheldon Vanauken's, *A Severe Mercy* (London: Bantam Books, 4th printing, 1981) – both of which involve suffering the death of an intensely loved spouse.

[21] Firestone, *The Dialectic of Sex: The Case for Feminist Revolution*, p. 209.

Love is being psychically wide-open to another. It is a situation of total emotional vulnerability.'

'Love between two equals would be an enrichment, each enlarging himself through the other: instead of being one, locked in the cell of himself with only his own experience and view, he could participate in the existence of another – an extra window on the world. This accounts for the bliss that successful lovers experience: lovers are temporarily freed from the burden of isolation that every individual bears.

But bliss in love is seldom the case: for every successful contemporary love experience, for every short period of enrichment ...'[22].

There does not seem to be any mention of a father's love but as regards the imperfection of a mother's love, 'we want to destroy this possessiveness along with its cultural reinforcements [of 'a mother who undergoes a nine-month pregnancy' and who is therefore ... 'likely to feel that the product of all that pain and discomfort "belongs" to her'] so that no one child will be *a priori* favoured over another, so that children will be loved for their own sake'[23]. Indeed, in Firestone's imagined 'households'[24] in which a group of people would live she comes as close as anywhere to recognizing the value of any type of family life, when she says: 'Adults and older children would take care of babies for as long as they needed it ... as in the extended family – no one person would be involuntarily stuck with it'[25]. Thus 'all relationships would be based on love alone, uncorrupted by dependencies and resulting class inequalities'[26].

---

[22] Firestone, *The Dialectic of Sex: The Case for Feminist Revolution*, p. 115.
[23] Firestone, *The Dialectic of Sex: The Case for Feminist Revolution*, p. 208.
[24] Firestone, *The Dialectic of Sex: The Case for Feminist Revolution*, pp. 206-216.
[25] Firestone, *The Dialectic of Sex: The Case for Feminist Revolution*, p. 209.
[26] Firestone, *The Dialectic of Sex: The Case for Feminist Revolution*, p. 209.

Love, in other words, is the implicit answer to the problem of pain; however, in the midst of pain there is an almost total eclipse of the possibility of love; and, therefore, love alone needs more than ever the suffusing presence of God: not a God unidentifiable with the problem of pain – but a God in whom pain is experienced fruitfully in suffering love to be lasting: indeed suffering love to be everlastingly good for the other.

Bioethical problems and the problem of love

The human experience expressed in the Scriptures, as we have seen, echoes in its complexity and difficulties, a whole range of modern bioethical problems; and, indeed, the reason for the present being echoed in the past is the depth of the desire to be fertile: to be with child. At the same time, however, even in the midst of all the difficulties and complexities of life there is the constant call to draw close to God and to experience the power of that word which fulfils the reality it prophesies: that "a man shall leave his father and mother and be joined to his wife, and the two shall become one" … "So they are no longer two but one. What therefore God has joined together, let no man put asunder" (Mt. 19: 5-6). In other words, whatever the situation we are in and the obstacles there are to its resolution, there is a power of God to bring us to the truth that sets us free (cf. Jn. 8: 32) to love. There are, however, many proposals to ignore the reality of human love as entailing an equal recognition of the dignity, gifts and differences between one man and one woman. Indeed, instead of embracing the call to conversion to rectify the imperfections of human love, there is an almost overwhelming possibility of these imperfections being compounded by the profound inability of adults, trying to solve the reality of sin without seeing it for what it is, such that they end up actually promoting it as a part of a misguided liberation from the social structures deriving from the marriage of a man and a woman. 'Then "the tyranny of the biological family would be broken," "unobstructed pansexuality" would replace heterosexuality, and "all forms of sexuality would be allowed and indulged." Firestone argued that "[u]nless revolution uproots the basic social organization, the biological family ... the tapeworm of exploitation will never be annihilated"[27]. In other words, out of all the confusion and pain concerning the nature, almost, of the patriarchal nuclear family, there seems to have arisen the whole disorientating ferment which wants to overthrow not just 'the biological family' but the whole structure of human sexuality – for if the roles are typically defined as the *robotic father* and the *unhappy mother* then what is there to choose from? Beginning, then, with the disorientation that arises out of a child being unable to identify with the same sex as his or her parent then there has arisen a kind of explosion of the sexes: that if what exists is so fundamentally disordered or unsatisfactory then it is necessary to invent a whole new range of possibilities. At the same time, however, there is a "naked use" of this confusion about identity to promote an anti-life programme: 'Many social scientists are now proposing as a solution to the population problem the encouragement of 'deviant life styles' that by definition imply nonfertility'[28].

How many bioethical problems, then, have arisen from the separation of 'the unitive significance and the procreative significance' (*Humanae Vitae*, 12) of marital love? There are the problems that arise out of not knowing who the parents of a child are, if egg and sperm are obtained from men and women and deposited in a "bank" – already communicating the whole encroachment of commercialism on the life of the person; and, indeed, there

---

[27] Excerpt from an article entitled, "The Trans-Industrial Complex" by Mary Hasson, *Humanum: Issues in Family, Culture and Science*, Issue 2: http://humanumreview.com/articles/the-trans-industrial-complex.The article has a link Firestone's book *The Dialectic of Sex: The Case for Feminist Revolution*:https://www.amazon.com/Dialectic-Sex-Case-Feminist-Revolution/dp/0374527873/ref=sr_1_1?ie=UTF8&qid=1536436757&sr=8-1&keywords=shulamith+firestone+the+dialectic+of+sex&dpID=514jT9a7qWL&preST=_SY291_BO1,204,203,200_QL40_&dpSrc=srch

[28] Firestone, *The Dialectic of Sex: The Case for Feminist Revolution*, p. 204.

are all the sufferings entailed in the "use" of women to provide "wombs" for other people's children or, in part, their own, depending on how many people are involved in the growing manipulation of "making a baby". Thus there is the whole "manufacturing mentality" which comes into play when technicians are directly involved in the transmission of life outside the womb, with all the processes of trial and error, testing, discarding and freezing who is unwanted. In other words, if there was a genuine recognition of each human life then there would be no deliberate loss of human life.

Thus it is possible to reject manhood, womanhood, marriage, family life and indeed the gift of human fertility and to waste the seed which makes it possible; and, in the process, to disappoint and to frustrate the desire for a child – both in the man who wastes his seed and in the woman who suffers this loss (cf. Gn. 38: 8-30). It would seem from the commentary on the text that Onan was taking advantage of the situation in that 'by marrying his sister-in-law he gets control over his dead brother's property – control which he can retain because he avoids having children by the woman'[29]. Whatever the ancient practice, then, for avoiding the conception of a child it has grown into a massive industry which reflects the same fundamental insensitivity as expressed in Onan's unwillingness to raise up children for his deceased brother's wife; and, ultimately, the hidden anti-life mentality becomes explicit in the modern practices of contraception. It is unlikely that the following reality is well known in that there seems to be such a blasé attitude to the existence of contraception[30]; indeed, like abortion, the reality of it is

---

[29] *The Navarre Bible: The Pentateuch*, p. 186.

[30] Firestone, *The Dialectic of Sex: The Case for Feminist Revolution*, p. 29, p. 185 and on p. 179 where she says: 'Present oral contraception is at only a primitive (faulty) stage, only one of many types of fertility control now under experiment' and then on p. 184: 'A feminist revolution could be the decisive factor in establishing a new ecological balance: attention drawn to the population explosion, a shifting emphasis from reproduction to contraception ...'. But to Firestone's credit, she recognizes that the 1970 'Pill Hearings' (footnote 1, p. 179) have just begun; and the

scarcely recognized. Thus there is the traumatic trampling of the poor, of justice, of truth, of scientific expertise and the suffering of others in the brutal pursuit of contraceptive substances: "Dr. Philippe Schepens says: 'Going back to the evening I spent with the Puertorican doctor, in addition to explaining all kinds of technical details of the experiments conducted by Pincus[31], he gave me a staggering account of the suffering of thousands of young women subjected to scandalous experiments, comparable to those conducted by Dr Mengele in Auschwitz. Because Pincus required an accurate assessment of the risks, he therefore administered hormonal doses more than a thousand times greater than the present dosage. The results were more than "convincing", because hundreds of women suffered the complications

---

'Pill Hearings' were 'investigative reporting of a pioneering journalist, Barbara Seaman, who uncovered and publicized the Pill's serious and life-threatening health risks': https://nwhn.org/pill-hearings/.

[31] In an article entitled "Guinea pigs or pioneers? How Puerto Rican women were used to test the birth control pill", by Theresa Vargas (May 19th, 2017: https://www.washingtonpost.com/news/retropolis/wp/2017/05/09/guinea-pigs-or-pioneers-how-puerto-rican-women-were-used-to-test-the-birth-control-pill/)

Gregory Pincus is described as the biologist by whom 'Puerto Rico was chosen as a testing ground' for the contraceptive pill because 'the location's overcrowding and poverty made it especially attractive to biologist Gregory Pincus, who was concerned about global population control.' But as the article also says, in the following words: 'Years later, following congressional hearings in 1970, some women would still question the pill's safety and the process behind the drug's approval. The Washington-based Women's Liberation group, issued a statement at the time, saying: "In spite of the fact that it is women who are taking the pill and taking the risks, it was legislators, the doctors, and the drug company's representatives, all men of course, who were testifying and dissecting women as if they were no more important that the laboratory animals they work with every day." The article does not address, however, the question of whether contraception is right or wrong and, according to a certain mentality of emotional persuasion, cites particular examples of the suffering of women to justify what was done.

Chapter Seven 253

indicated in the leaflets in the "pill" boxes'"[32]. How, then, can a product that embodies the mentality that is prepared to destroy the lives of women be claimed to be of help to them?

How many bioethical problems have arisen from being unable to recognize the reality of human personhood being from the first instant of human fertilization until natural death; and, indeed, is the reason for being unable to recognize the reality of the origin of human personhood in the first instant of fertilization because it means recognizing the personal existence of the subject? In other words, if the humanity of the personal existence of the subject is recognized from the first instant of fertilization, then how many practices and procedures are immediately seen to be the unethical pursuits that they are? So called embryo stem cells cease to be an "experimental raw material" and are recognized to be what they are: cells integral to the life of the developing child. Where, in other words, is the dividing line between being conceived and being human if twinning, actually, brings about two human lives from the one conception? Whether, then, one, two, three or more children are conceived from one conception is irrelevant to recognizing the humanity of them all and indeed marvelling at what actually occurs. The real problem, in the end, is the simple recognition of human life expressing human personhood[33]; and, in the end, what would be the point of all these procedures, freezing of human embryos – if they did not involve the prized priceless gift of human life?!

At the same time, too, if the question of meaning is integral to human life

---

[32] "*Humanae Vitae*: My testimony as a doctor" by Philippe Schepens, General Secretary of the World Federation of Doctors who Respect Human Life, delivered at "*Humanae Vitae* at 50: Setting the Context", Pontifical University of St Thomas Aquinas, Rome, 28 Oct 2017: http://voiceofthefamily.com/dr-phillippe-schepens-humanae-vitae-a-medical-doctors-testimony/.

[33] Cf. Francis Etheredge, *Conception: An Icon of the Beginning*, from enroutebooksandmedia, 2019:

http://enroutebooksandmedia.com/conception/.

and suffering it is essential that at whatever time of life there are sufferings that the question of its meaning is not overlooked, disregarded or ignored. The desire to die arises in the human heart in the most profound trials of life. However, the desire to die expresses, in its own way, the insufficiency of each of us to suffer alone; indeed, it is like living as if there is no oxygen if we live without seeking out the meaning of human sufferings. In other words, just as the very origin of life arises out of the reciprocal gift of love between husband and wife[34], so the very challenges of life are to be lived in dialogue with others: in the "being with others" that expresses the mystery of love beyond death.

In the end, then, we need the whole truth about human being and not scraps of suffering to attract our sympathy and distract us from the reality of what is being done, or ideological trampolining displays of partial truths to win our admiration while embryonic children are frozen out of sight or subjected to all kinds of quality control tests. We need, finally, a redeeming love which rescues, corrects, teaches, helps and makes possible the perfection of human love. We need, in other words, God the Father's answering Love, conceiving the Son of God through the power of the Holy Spirit, welcomed and borne by Mary, spouse of St. Joseph, together rearing the One who will restore the family through the Holy Family of Nazareth, that we may see more clearly the Love that God Is.

---

[34] A book the author is currently working on is provisionally called, *Human Nature: Natural Norm*.

Epilogue

In the opening prologue it seemed as if there is a growing rejection of the reality of womanhood, motherhood, manhood, fatherhood, of a family with children, of a world capable of bearing the gift of human life and sharing resources, of the meaning of suffering and of love; indeed, the ultimate root cause of it all was a psychology of power which, it was claimed, originated with the distinction between being male and female. Without the dynamic of sin and salvation, then, men and women seek an analysis in which a cause of the present pain is proposed and, as it were, universalized. With the mentality of blaming the 'biological family', 'reproduction' and a 'rigid patriarchal' oppression of women and children, there is the advance of an answering rejection of these realities – a rejection which, in one sense, seems justified on the basis of crippling human experience. On the other hand, however, there seems to be barely any recognition of the positive experience of the unfolding of human personhood; and, therefore, there seems to be an increasing radicalism which, rather than going to the root of the problem of pain, increasingly overthrows anything that resembles the cause of it.

The manufacture of children and the reality of relationship

In terms of bioethics, then, there is an uncritical, almost unconscious, movement towards the manufacture of children as if the resolution of suffering oppression is to inflict it in as dehumanizing a way as it has been experienced; indeed, to even think it possible that there can be a manufacture of children is already a measure of the extent to which people are alienated from the everyday gift and task of fostering the relationships which spring from marriage. It is almost as if the "production process" expresses, in a way, how far the stripping away of the relational nature of human life has already progressed; and, in a sense, the whole confusion about sexual identity expresses the same thing: that just as a person does not think in terms of being-in-relationship so that person cannot identify him or herself as a man or a woman. What we see in Firestone's observation of the desirability of manufacturing is a mixture of justified sympathy for the woman who bears the child and, at the same time, an extraordinary lack of empathy for the child who bears the brunt of this rejection – not dissimilar to the rejection that she has experienced and written about, directly and indirectly, throughout her book:

> 'Child-*bearing* could be taken over by technology, and if this proved too much against our past tradition and psychic structure (as it certainly would at first) then adequate incentives and compensations would have to be developed ... to reward women for their special social contribution of pregnancy and childbirth'[1]. Furthermore, Firestone says: 'Any child-rearing responsibility left would be diffused to include men and other children equally with women'[2].

In terms of a relationship to children conceived through a process of manufacture, Firestone does not understand that it is not just a matter of it being 'too much against our past tradition and psychic structure' – as if these were surmountable obstacles like westerners eating insects out of solidarity with the poor and to benefit our environmental home. In other words, listening to the suffering of a woman who, desperate to conceive, agreed to an *in-vitro* fertilization program because the evidence suggested that she was not going to conceive naturally, although she did, strikes at the heart of this misconception about the humanity of mother and child (the father is scarcely

---

[1] Firestone, *The Dialectic of Sex: The Case for Feminist Revolution*, p. 213.
[2] Firestone, *The Dialectic of Sex: The Case for Feminist Revolution*, p. 214.

mentioned). In view, then, of the *in-vitro* fertilization technicians' incentive of free "treatment", she agreed to the fertilization of eight children, one of whom so far has been born, five of whom have died and two of whom are still frozen in liquid nitrogen with all the angst and costs that these procedures have subsequently entailed. "Katy" says, among the many other things mentioned above, including regret: 'I knew I would need to have them all because despite my egregious disregard of Church law in doing IVF at all, I still fervently believe that life begins at conception and that those … little souls would absolutely not be destroyed or donated to science'[3]. In other words, there is a kind of conversion to the reality of being "mother and child" and suffering through what has been done (presumably with the cooperation of the father); and, therefore, the reality of a relationship to real children and their growth goes beyond the abstract concepts of child manufacture and all that that entails.

At the same time as imagining, like many others who are now doing it, that the manufacture of children can be good, Firestone recognizes that there is a real need to help the mother herself and for there to be help in the rearing of children. On the one hand there is the need for 'adequate incentives and compensations' to help women in their child-bearing years; but, in addition, there needs to be the involvement of men, the husbands and fathers one could argue, of the children themselves, and for other children to be involved in child-care too. In other words, put positively, the extended family is a healthier society than the proverbial patriarchal nuclear family. Indeed, in a recent, extended review of the family in America, the author notes: 'People who grow up in a nuclear family tend to have a more individualistic mind-set than people who grow up in a multigenerational extended clan'; and, what is more, it is not as if the multigenerational family consists

---

[3] "'Katy': IVF Regrets: One Mother's Story", p. 3 of pp. 2-3 of the Fall Newsletter, Issue 4, 2019, "tiny sparks":

https://sacredheartguardians.org/wp-content/uploads/2019/09/Tiny-Sparks-2019-Fall.pdf (source of reprinting given in the aforementioned citation).

solely of relatives: 'Kinsmen belong to one another, Sahlins writes, because they see themselves as "members of one another'[4]. In a word: love needs to go beyond ourselves and to benefit others in a wide range of ways.

On the other hand, the involvement of technology which is substitutive of the personal nature of the human conception of children already implies a depersonalization which is contrary to the very good of the parents; and, at the same time, the depersonalization of the production of children is like trying to strip the bark off trees without killing it: it is an action contrary to the very nature of the human sensitivity necessary for normal development. The complete and utter irony of a rejection of oppression by one group of people involving the complete subjection of others is an indescribable witness to the inhumanity which comes to be present through the corruption of power.

Biblical prophecy and the Church's criticism of fathers

Biblically, however, the problem of power in the relationship between the sexes is itself a consequence of original sin; indeed, recognizing what has changed because of sin, God says to Eve 'your desire shall be for your husband, and he shall rule over you' (Gn. 3: 16). But, at the same time, the restoration of a good relationship between the two sexes is implicit in the prophetic promise to the serpent of the woman's seed which will crush his head: "I will put enmity between you and the woman, and between your seed her seed; he shall bruise your head, and you shall bruise his heel" (Gn. 3: 15).

---

[4] In this excellent and wide ranging study of the American family over the last few generations there is much to commend it and much to help people reflect on the value of the extended family that is open to the needs of others: David Brooks, "The Nuclear Family Was a Mistake: The family structure we've held up as the cultural ideal for the past half-century has been a catastrophe for many. It's time to figure out better ways to live together": theatlantic.com/magazine/archive/2020/03/the-nuclear-family-was-a-mistake/605536/.

As the wisdom of God prophesied and then brought about the covenantal renewal of marriage with the coming of Christ, so the Church took up the work of correcting men and women's excesses; however, recognizing a particular problem in 1880, Pope Leo XIII says in *Arcanum*: it is necessary to limit the 'power of fathers of families, so that sons and daughters ... are not in any way deprived of their rightful freedom' (15; and cf. also 26)[5]. In other words, there is an ongoing work of addressing the sin at the basis of human imperfection, particularly the imperfections in the love of husbands for their wives. Indeed, 'the whole contraceptive mentality appeals, in a sense, to an escape from the responsibility of the man for his child'[6]; and as regards those who 'decide upon the death of the child in the womb ... [in the first instance] the father of the child may be to blame' (St. John Paul II, *Evangelium Vitae*, 59).

Finally, in *Christus Vivit*, the Post-Synodal Apostolic Exhortation following the Synod on the Youth, Pope Francis wrote: 'Perhaps your experience of fatherhood has not been the best. Your earthly father may have been distant or absent, or harsh and domineering. Or maybe he was just not the father you needed. I don't know. But what I can tell you, with absolute certainty, is that you can find security in the embrace of your heavenly Father, of the God who first gave you life and continues to give it to you at every moment. He will be your firm support, but you will also realize that he fully respects your freedom'[7].

---

[5] Cf. Quoted by Etheredge, on p. 57 of *Scripture: A Unique Word*, 2014; and cf. the whole of "Chapter Two" of that book for a more extensive discussion on the vocation of the father.

[6] *Scripture: A Unique Word*, Etheredge drawing on what Bishop Karol Wojtyla (later St. John Paul II) said in his spiritual counsel to male students p. 56 of the *Way to Christ: Spiritual Exercises*, translated by Leslie Wearne, HarperSanFrancisco: A Division of HarperCollinsPublishers, 1994.

[7] *Christus Vivit,*

Is there a way back to the future of marriage and family life?

'In his general audience of December 19, 2012, Pope Benedict XVI gave this commentary on the Annunciation:

"Today, I wish to ponder briefly with you on Mary's faith, starting from the great mystery of the Annunciation.

Chaîre kecharitomene, ho Kyrios meta sou, "Hail, [rejoice] full of grace, the Lord is with you" (Lk 1:28) ... The Angel's greeting to Mary is therefore an invitation to joy, deep joy. It announces an end to the sadness that exists in the world because of life's limitations, suffering, death, wickedness, in all that seems to block out the light of the divine goodness. It is a greeting that marks the beginning of the Gospel, the Good News"'[8].

In front of the possibility of Mary becoming the Mother of the Lord 'She can only say, "How shall this be?" God is clearly pleased with this humble response, which in reality is a petition for the Power of the Most High. She desires precisely what God intends to bring about: she will be overshadowed by the Holy Spirit'[9].

'In effect, the Holy Family experienced such starkly cold surroundings and environment of oppression that they faced the challenge of developing and maturing in this context. In the end, Mary becomes the

---

http://w2.vatican.va/content/francesco/en/apost_exhortations/documents/papa-francesco_esortazione-ap_20190325_christus-vivit.html.

[8] Taken from "A Moment with Mary", who translated the excerpt from an article on Zenit by Anita Bourdin, December 19, 2012 and posted at: https://us3.campaign-archive.com/?e=83d33a4ae4&u=bbaf519c73482457368060b5b&id=1d54d6e03b.

[9] Douglas G. Bushman, "'How Shall This Be?' A Call to Humility for a New Pentecost': https://www.hprweb.com/2019/03/how-shall-this-be/.

clearest, purely human North Star, or reference point, for these circumstances'[10].

When so many have already become witnesses of the culture of death[11], when so many have lost their way with their human identity and when there is such a powerful program planning the selective suppression or alteration of the human race – where is it possible to go from here except further down the path of deconstructing the human race in an endeavour to frustrate population growth? To go forward is not about forgetting the past and its progress from principle to practice and from practice to principle; to go forward is about going back to a beginning that founds the future on what experience has already shown to be enduring. To be without the past is to be trapped in the present that proceeds from it; to recognize the present expression of the past is to recover the roots that need to be renewed if the promise of the future is to become a real possibility. Thus, as Pope Francis says: 'That is how various ideologies operate: they destroy (or deconstruct) all differences so that they can reign unopposed. To do so, however, they need young people who have no use for history, who spurn the spiritual and human riches inherited from past generations, and are ignorant of everything that came before them'[12]. But to be aware of what 'came before them' does not only belong to the immediate past of a present program;

---

[10] Deacon Thomas Baca, "Mary the Perfected Witness: Achieving Our Saintly Destiny in a World of Choices":
https://www.hprweb.com/2019/03/mary-the-perfected-witness/.

[11] Cf. Francis Etheredge, *The Prayerful Kiss*, 2019: https://enroutebooksandmedia.com/theprayerfulkiss/, for prose and poem on the death of a child of mine from abortion; and cf. Etheredge, *Conception: An Icon of the Beginning*, 2019:
http://enroutebooksandmedia.com/conception/, for an in-depth investigation of the mystery and meaning of human conception.

[12] Pope Francis, *Christus Vivit*, 181.

rather, it is to tune into the great themes which engage in us the breadth and depth of the search for truth: the truth that is always present in the reality of "today".

Why, then, begin the last part of the book with a return to Mary? In other words, there is a kind of impasse in the mentality of the day; indeed, even if it is possible to recognize many true criticisms of men, there seems to be a mentality abroad which is insensitive to the true good of marriage, family life and the contribution of men to society. In other words, it is not so much that certain criticisms are not justified but that there is the assumption[13] that there is no good beyond the justified criticisms: that marriage and family is irretrievably, irremediably and irrevocably beyond remedy. Perhaps the answer lies in a certain mentality: humility.

---

[13] Cf. Frederik Andersen, Rani Lill Anjum and Elena Rocca, "Philosophy of Biology: Philosophical bias is the one bias that science cannot avoid", on p. 2 of 5, they say: 'Biology, for example, is concerned with both entities and processes (Nicholson and Dupre, 2018). The standard ontological assumption is that entities (such as proteins) are more fundamental than processes, and that processes are produced by interacting entities. Molecular biologists have traditionally taken this as the default position. The ability of entities, such as proteins, to interact with each other is determined by their chemical structure, so to understand processes (such as interactions between proteins), we need to understand the entities themselves in detail.

However, some scientists take the view that processes are more fundamental than entities (Guttinger, 2018). In this view, entities are understood as being the result of processes that are stable over some length of time, and the best way to understand the behaviour of an entity is to study the relations it has with other entities, rather than its internal structure. Ecologists tend to take this view, thinking in terms of systems in which the properties of individuals and species are determined by their relationship with each other and their environment':
https://www.academia.edu/38645400/Philosophy_of_Biology_Philosophical_bias_is_the_one_bias_that_science_cannot_avoid.

Reasonable and graced humility

On the one hand there is, as it were, the possibility of reason. There is the fact that the existence of the complementary difference between men and women is personalistically coherent; indeed, the very equality of persons is established by each being both indispensable to the other and different: the complementary difference being what each brings to the relationship between the sexes. This is especially evident in marriage and the founding of a family; indeed, it is one of the personalistic characteristics of the equality of marriage that it entails one man and one woman. The perfectibility of marriage is a work which, according to time and circumstance, is an ongoing work; indeed, dysfunctional ways of relating to each other or imperfections prejudicial to equality between the sexes are a constant stimulus to the challenge of drawing on the complementary whole of each sex. At the same time, however, it is unrealistic to imagine that deconstructing the sexes would "cure" the fundamental problem between them; rather, deconstructing the sexes would lead to innumerable problems, not least of which are both the obscuring of each person's identity and the danger of this mentality becoming an imposition on others.

The overriding assumption that the cultural contribution to gender differences makes roles flexible to the point that a man or a woman's body can be manipulated to the extent of surgical, hormonal or even genetic changes being superimposed on an existing sex is a profoundly theoretical claim[14]; indeed, this is totally different to the predicament of an ambiguous bodily sex which requires all the sensitivity, discernment and specialist knowledge available to be of service to the person in need of help. One of the tragic ironies of programs of reform is the danger of oppressing others with

---

[14] Cf. Edwin Benson, "Why Transgender Equality is the Death of Women's Sports": http://www.returntoorder.org/2019/03/why-transgender-equality-is-the-death-of-womens-sports/?PKG=RTOE0651.

even worse suffering than those which have already been experienced. Consider the tragic exploitation of people under totalitarian regimes as well as the almost global attack on the fertility of the peoples of the earth and, in the process, the untold death of those conceived and aborted, frozen or exploited for research; for if, as even the Warnock Report concluded, there is no obvious point of humanization once conception has begun, then it follows that the whole process is integral to the unfolding of the person from conception: "there is no particular part of the developmental process that is more important than another; all are part of a continuous process"[15]. Thus there is an evidence-based rationality which recognizes that the manifestation of the person is a process which unfolds what was present from the beginning. Humility, in this instance, is being willing to admit that the evidence determines the possibility of a person being present from conception; and, indeed, corresponding to the original meaning of the word, this is an attitude that proceeds, as it were, from the facts of the 'ground'[16] of human identity.

On the other hand there is the possibility of Grace: 'In God's wisdom, the apostles' humility over confronting their inability to follow Jesus without the Holy Spirit is the ultimate disposition for receiving the Holy Spirit'[17]. Mary, however, 'is simply aware that she needs God's grace …. God is clearly pleased with this humble response, which in reality is a petition for the Power of the Most High. She desires precisely what God intends to bring about: she will be overshadowed by the Holy Spirit[18]. Or in the words of Pope Francis:

---

[15] Department of Health and Social Security (UK), *Report of the Committee of Inquiry into Human Fertilisation and Embryology* (London: Her Majesty's Stationery Office, July 1984), para. 11.19, quoted in Catholic Bishops' Joint Committee on Bio-ethical Issues, *Response to the Warnock Report on Human Fertilisation and Embryology* (London: Catholic Media Offices, 1984), 13.

[16] https://www.etymonline.com/word/humility.

[17] Bushman, "How Shall This Be?" A Call to Humility for a New Pentecost'.

[18] Bushman, "How Shall This Be?" A Call to Humility for a New Pentecost'.

"We must remember that prayerful discernment has to be born of an openness to listening – to the Lord and to others, and to reality itself, which always challenges us in new ways. Only if we are prepared to listen, do we have the freedom to set aside our own partial or insufficient ideas... In this way, we become truly open to accepting a call that can shatter our security, but lead us to a better life. It is not enough that everything be calm and peaceful. God may be offering us something more, but in our comfortable inadvertence, we do not recognize it"[19].

The radical originality of God

Returning to Mary, then, is about rediscovering, renewing and appreciating afresh an anthropology of gift: the gift of being ourselves and beginning with where we are – but especially the gift of being loved unconditionally with an eternal love. Love begets love; and, therefore, it is in believing that we are loved that God goes to the depths of our reality and transforms us in ways beyond purely human potential – but which open upon profoundly new human possibilities. The Virgin Mary, then, experienced 'the power of the Most High' (Lk. 1: 35): the power that brings what is radically new to exist. On the one hand there is the possibility of being like the person described by Venerable Carlo Acutis when he said: "everyone is born as an original, but many people end up dying as photocopies"[20]; and, therefore, without realizing it, we become the repetition of what others have written: uncritically reproducing what may even be partly true. On the other hand, 'the power of the Most High' promises a renewal of the truth that takes up what is both ancient and modern and remakes the synthesis that speaks beautifully of the coherent whole: 'Let us make man in our image, after our

---

[19] *Christus Vivit*, 284, citing a quote from the Apostolic Exhortation <u>Gaudete et Exsultate</u> (19 March 2018), 172.

[20] Quoted by Pope Francis in *Christus Vivit*, 106.

likeness .... So God created man in his own image, in the image of God he created him; male and female he created them' (Gn. 1: 26, 27).

What awaits us, however, is the full answer to the question of why we are made man, male and female. Indeed, there may well be an unimaginable answer to that question as, God willing, we enter eternity and discover, in a blazingly amazing way, that "we" are filled with an illuminating brilliance that can only be described as being a befitting expression of the mystery of the Blessed Trinity![21]

---

[21] Cf. Kimberly Bruce, *The Gender Link to the Human Soul*, forthcoming from *En Route Books and Media*, 2020; endorsed by Francis Etheredge: 'In this brief study Kimberly Bruce has compiled a comprehensive account of the biological, psychological, sociological, Scriptural and Magisterial sources on the abiding reality of embodied personhood, drawing particularly on the living mysteries of the Son of God, Jesus Christ and His Mother Mary; and, as such, leads to the possibility of enriching a growing body of reflection and stimulating further work on the significance of being a male-person or a female-person.

The eternally enduring nature of being male and female raises the following, blazing questions: What is the everlasting significance of being male and female? What is the resplendent revelation of the person-gift-sign of being either a man or a woman which awaits us in heaven? In what way will the mystery of the Blessed Trinity brilliantly illuminate the fact that man and woman are made in the image and likeness of God?'

An End Word

A New Beginning
Bishop John Keenan of Paisley,
Scotland

I take up this endnote from words with which Francis Etheredge brings this refreshingly imaginative work of bio-ethics to a close. 'To go forward is about going back to the beginning'. Of course, the Beginning we mean is not what common parlance talks of when it speaks about a start. The Beginning we mean is the same thing as Creation. It is an event *ex nihilo* and so without precedent in the most literal sense.

Ex nihilo means beginning from nothing only when taken in a sort of negative sense. More positively, and more realistically, it tries to comprehend an act that contains within itself every originality, and with all possibility at its disposal. Its fruit is, in consequence, unique and unrepeatable; by definition extraordinary.

Neither can the ultimate source of such a Creation be exhausted in reason or science, even if the reality that unfolds from it will contain the fullest meaning and make more sense than could ever have been anticipated. Reason cannot offer Creation's ultimate explanation because reason, reflected into the practical sphere, involves an obligation to something prior, something already given.

Only love comes close to being its own explanation, from our way of seeing things, and is, therefore, entirely free. While what reason produces is owed in justice, what love creates is somehow pure grace and gift, unmerited and simply to be received with wonder and appreciation. Venerable Carlo

Acutis puts his finger on it when, revealing how our world is an original, sets the perennial human challenge not to treat its fruit as if it were a mere photocopy.

Mary of Nazareth, in the Angel's nomination of Her as *gratia plena* and similarly awaiting Her hoped for *fiat*, knew then the unfathomable depths of Her being and person as a co-principle of Creation. In Her we discover the fuller mystery of humanity's vocation to cooperative partnership with the Creator. Nor is our *fiat* to God's creative act simply an affirmative but, in the power of the Holy Spirit, is transformed and exalted into a pro-creativity. In the end, Mary's 'Yes' to all that is from God, *usque ad finem*, allows God's Creation to go on until Her Son, in Whom all that is came to be, presents it at the last to the Father, All in All.

All of this seems to me the perspective of Francis Etheredge's unique contribution to the Christian vision of human bio-ethics. Revelation, and particular the person of Mary, help uncover to bio-ethics its only worthy starting point, namely awe and appreciation before the original gift of the procreation of the human person.

That is why Etheredge begins with something as unexpected as the Holy Family of Nazareth. It opens up to us the logic of gift. Mary and Joseph long to bear in their union the full meaning of the Old Testament gift of covenant between God and humanity, manifested in their marital pact. Their union is no merely human contract or transaction with the logic of *quid pro quo* and always concealing some 'if or but'. Uniquely their marriage is an interweave of unconditioned promises, given freely and gladly in love as a gift of self, and received in gratitude with the reciprocal response of self.

As if consequently, it is in the everyday experience of family life, each member unique but all finding themselves in each other, appreciating the gift of the other while bearing each other's burdens, that we discover our authentic personality in relation to personal love received somehow unexpectedly and returned gratefully.

An End Word

The person of Mary grounds all these originalities. She knows Her identity is founded in being chosen by God for no other reason than love. Since all depends on His grace rather than Her natural ability, there is always reason to hope because the human past, present and future are in His hands, not our possibilities. So, Her first act in Luke's account is not to produce, as if all depended on Her, but to accept God's good will in prayerful dialogue in the realisation that all comes from Her Father God, Who is Her Creator.

Mary's logic is the fruit of the realisation of Her origins. She knows interiorly how there was never a moment when anything of Her substance existed without it being loved as personal. There was no conception that paved the way for Her. Instead, Mary, as an original work, came to be as the beloved conceiving of a person, unique because incomparably loved by Her Father.

As ever, there is so much that is original in Etheredge's work because he begins from a place no-one ever imagined, such that all that follows is somehow novel. This endnote cannot begin to comprehend all of its parts.

So, if I may content myself with this following and leave it there. That is, from such speculations Etheredge reasonably raises the question about whether the dogmatic definition of the Immaculate Conception does not collaterally define the moment of the ensoulment of every person as at conception. He certainly raises a case to answer.

From what has been noted above, of course, the implications of the Church answering in the affirmative would not only be to settle the moral dilemmas inherent in this bio-ethical question.

For Etheredge it would be to reset the first principles of bio-ethics itself as a whole, moving on this neonate discipline from currently vexed questions around how and where it might start as a science, and opening it onto the awesome horizon of that mystery where it will always and only find its real Beginning.

www.ingramcontent.com/pod-product-compliance
Lightning Source LLC
Chambersburg PA
CBHW031315160426
43196CB00007B/535